BETWEEN WOMEN
AND GENERATIONS

BETWEEN WOMEN
AND GENERATIONS

Legacies of Dignity

DRUCILLA CORNELL

palgrave

BETWEEN WOMEN AND GENERATIONS
Copyright © Drucilla Cornell, 2002.

Softcover reprint of the hardcover 1st edition 2002 978-0-312-29430-4

First published 2002 by PALGRAVE™
175 Fifth Avenue, New York, N.Y. 10010 and
Companies and representatives throughout the world.

PALGRAVE is the new global publishing imprint of St. Martin's Press LLC
Scholarly and Reference Division and Palgrave Publishers Ltd (formerly
Macmillan Press Ltd).

ISBN 978-1-349-63488-0   ISBN 978-1-137-09871-2 (eBook)
DOI 10.1007/978-1-137-09870-2

Library of Congress Cataloging-in-Publication Data
is available from the Library of Congress.

Design by Letra Libre, Inc.

First Palgrave edition: May 2002
10 9 8 7 6 5 4 3 2 1

As promised
to my mother
Barbara June Kellow Cornell
November 27, 1925–August 25, 1998

CONTENTS

Freeing yourself was one thing;
claiming ownership of that freed self
was another.

—*Toni Morrison,* Beloved

ACKNOWLEDGMENTS

Given the difficulty of writing a book that fulfills a promise to my mother, it has been a blessing to me to find people who could support me in this endeavor. As I tried to stay true to my promise, many crucial and special people came forward to help me along the way and make the completion of this book possible.

I want to thank the innumerable participants in seminars and lectures who helped me develop the ideas in this book. Their questions and insightful comments challenged me to reevaluate the ways in which my own Eurocentrism can lead me astray. I have learned more than I can say. I owe it to peers, colleagues and students who, from within their own cultural perspectives, were willing to engage with me in their pursuit of lessons about the ethics of dialogue and witnessing—lessons on which this very book centers.

Roger Berkowitz and Sara Murphy gave me invaluable support throughout the writing of this book. Each of them took the time to read the entire manuscript and to reread each of the chapters after they were revised in accordance with their suggestions. Such committed friends and interlocutors have been crucial in helping me tackle the difficult questions of style, approach, and method this book presented. I want to thank Judith Feher Gurewich, Graciela Abelin Sas, Doris Silverman, Jessica Benjamin, and all other practicing analysts who provided their insightful comments and suggestions for chapter 2. With his vast knowledge of psychoanalytic literature, Charles Shepherdson helped me clarify certain key concepts. I thank them all for the time they gave to carefully read the manuscript. Chapter 2 was profoundly enriched by their insights. I want to thank Judith Butler for her emotional and intellectual support of this project from its inception. As a friend as well as an intellectual interlocutor, Judith's role in my life has been priceless for the development of my work over a period of almost twenty years now. She gave me thoughtful comments on chapter 3 that allowed me to be more precise in some of the theoretical and conceptual distinctions I was trying to make in my reading of Immanuel Kant's *Critique of Judgment*. Through her own daring engagement with the intricacies and complexity of feminist history, Gayatri Spivak took me down an unexpected path and inspired to me to write chapter 3. She, too, has been a friend and interlocutor for close to twenty years, and her courage and brilliance have always pushed me in new directions. Gayatri also took the time to give me her own thoughts and comments on chapter 3, and the published text is undoubtedly much better because of them.

Cindy Crumrine was my copyeditor for *At the Heart of Freedom* at Princeton University Press. She was so helpful to me during that project that I asked her to edit the first two chapters of *Between Women and Generations: Legacies of Dignity*. Cindy truly walks an extra mile for the books she works on, and I have learned a great deal from her about what makes a sentence readable. I thank her for her efforts on both books. Karen Wolny, my first editor at St. Martin's Press, pursued this book with great enthusiasm. Her comments on the entire manuscript improved the book considerably. Sadly for me, though of course wonderful for her, Karen received an important promotion toward the end of the writing of this book. But she saw to it that I was given an excellent replacement editor, Gayatri Patnaik. Even though she was officially off the project, Karen read the last versions of the manuscript in her own free time and commented on it carefully. Her enthusiasm for the book, and her relentless insistence that it reach as wide an audience as possible, prodded me to reshape the presentation of some of my most essential ideas. She intellectually supported me. And by sharing with me its emotional impact on her, Karen reminded me that this book carries with it an important message. Karen went beyond what anyone would expect from an editor. I thank her for the diligence, energy, and tireless devotion she put into this project. Her voice echoes throughout the manuscript. Gayatri Patnaik took over with the same enthusiasm that I came to expect from Karen. Like Karen, she commented on the manuscript as a whole and guided the book through to its completion.

Michelle Zamora was my assistant at the beginning of this project, and continued to play a role in the organization of its completion. Both through our discussions and through the constant supply of reading material, Michelle inspired me to think in new ways. By helping me keep up to date with the writings of some of the most brilliant thinkers in Chicana and Postcolonial Studies, she kept me from going stale. She is an insightful and tough critic, but gracious enough so that whenever she catches me slipping up, it is never with a tone of righteousness or condemnation. She is someone I have learned from and from whom I will continue to learn as she now resumes her own journey into her academic career.

Dinh Tran committed himself to the project of helping correct my endless computer "screw ups" with footnotes. My enthusiasm for rewriting often led to dreadful consequences in the actual maintenance of the document. When confronting "fried" sections, which I sometimes haphazardly dropped from the body of the text into the footnotes, Dinh was always cheerful. As a Vietnamese refugee, he brought his own experience to bear on my engagement with the women of the housecleaners cooperative *Unity*. Dinh read every page of the manuscript and provided insightful comments. At times, his ideas concerning some of the philosophical concepts in this book actually changed my thoughts about them. During the course of his work with me, he befriended my daughter and helped us both think through the problem of how a minority child is to deal with the endless, small, and big insults that come with living in a racist society such as our own. Dinh has helped me remain "cool" in my daughter's eyes by keeping me up to date with the latest speech of young people. Writing a book is both a wonderful and frustrating endeavor. Dinh was vigilant in his reminders that no matter how frustrating things get, I should not "throw shade"—neither at him, nor at anyone with whom I might be working.

It has been a privilege over the last few years to have Dan Morris as a student. At the age of twenty-two, he is already an accomplished scholar with many publications. His knowledge of philosophy in general and critical theory in particular has been an endless and important resource for me. Dan and I have spent hours discussing the relationship between Romantic and Critical Idealism. Carefully and closely, we read Kant's *Critique of Judgment* together. He helped me shape my own reading of this particular work of Kant's. Dan read every page of the manuscript, and took the time to make intricate editorial suggestions, all of which have been included.

Constanza Morales-Mair invested both her intellect and her heart in the writing of this book. She was constantly at my side, editing and reworking the book sentence by sentence. Her insight and extraordinary writing skills improved the readability of this book enormously. Her emotional support for what undoubtedly became the most difficult book for me to write was as important as her incisive intellectual and critical editorial comments. Day after day, Constanza was not only by my side in the writing of this book; she also participated in all of the conversations and interviews with *Unity*. Indeed, it is no exaggeration to say that, without Constanza, I believe that the door to the kind of engagement we both ultimately had with the women in *Unity* would not have been open. Throughout the interviews, Constanza interpreted for me, since my Spanish was terrible and, to this day, remains, well, pretty bad. We both attended the parties and celebrations that took place as these interviews came to fruition. Constanza is a beloved friend and her wisdom has taught me more than I can say. This book reflects her spirit in almost every page. I thank her for her devotion and for her support; without it, I am not sure I would have been able to undertake this project and bring it to its fulfillment.

I want to thank all the women I interviewed in *Unity Housecleaners:* Zonia Villanueva, Zoila Rodriguez, Mónica Díaz, Antonieta, and M. E. A. I went in with one set of expectations—to interview a "nannies" cooperative—only to find out that all these women who had worked as "nannies" believe this kind of job is in and of itself exploitative. This is just one example among many of how these interviews unsettled my own convictions and made me aware of how many things in my own life remain unexplored. Chapter 4 is a tribute to them—to how they made me rethink so many things about my own life as a mother. Their courage and their commitment to make this world better for all women immigrants is exemplary. *Unity* shows just how much can be accomplished by a determined group of women, even when all the odds are against them. Mónica Díaz translated the first draft of Antonieta's interview. She also set up all the appointments for Constanza and myself. What started out as interviews developed into friendships that have made me see the world with new eyes.

My mother belonged to a group of bridge life-masters. A life-master in bridge is a person who has accumulated a large number of points through placing in or winning bridge tournaments. My mother played in bridge tournaments throughout the world since she was very young. In her last years she studied with a group of women, Bee Bayer, Sally Jac Schafer, and Bonnie Sakamoto. It was these women my mother chose to be the final arbiters of whether I had written an accessible book, and drawn a vivid and fair portrait of the woman they all loved and respected so much. My mother told them that her life and her legacy were to be put in my hands. I thank them all for reading the manuscript and for giving me their comments. I particularly want to thank Bonnie Sakamoto who read the entire manuscript twice and shared with me words that went straight to my heart: that this book brings my mother to life in all her complexity and struggle. She felt that I had done what my mother had wanted most for her daughter, the writer; to make the reader respect her life and her courage in death. I thank Bonnie Sakamoto for her support because, in the end, I knew I would not have carried out my

mother's desire if she had not found in this book the Barbara June Kellow
Cornell she knew. I cherish them all for the care they gave to my mother, and
to this story of who she was.

Since Sarita was a baby, Larry Brassell—Uncle Larry—has been part of
my life as well as my daughter's. He has accompanied me on all of my lecture
trips outside the United States. As my "adopted" brother, he has been unfail-
ing in his support for my work. I have learned a lot from him about how to
deal with the complexities of parenting. He has helped me through both the
best and the worst times. Without him, I cannot imagine how I would have
been able to continue my professional life and still be a mother in the ways I
had always hoped. Irena Molitoris has also been with Sarita and me since
Sarita was a baby. I have known Irena since she was sixteen, and I have had
the extraordinary pleasure of watching her grow up. My daughter calls Irena
her "sister" and she, like Uncle Larry, has joined me in the pursuit of creating
an extended family. Their devotion not only has given me the necessary support so that I can work and write but has also brought so much joy into our
lives. Irena is graced with the patience of a saint, and at times that was exactly
what she needed to have in caring for Sarita. Their love will always be part of
Sarita's story and of mine.

My daughter Sarita has already placed ethical restrictions on what I may
and may not say about her. But we have come to an agreement: I am allowed
to communicate my experience of and feelings for her. She has taught me
how vulnerable our love makes us. The stories of mothers and daughters
woven throughout this book are inseparable from my own reflections of who
I have become with her and through her. I thank her for the constant pressure to remain honest and in touch with our legacies.

To my grandmother, Mildred Kellow, I owe the images that often guide
me in my relationship with my daughter. My love for Nana is one of the miracles of my life. This book is undoubtedly part of keeping her spirit alive. Adventurer that she was, she left the paths she did not complete open for me to
pursue. For opening me up to new worlds and supporting me in my explo-
ration of them, I particularly would like to thank M. Jacqui Alexander. And in
the end, I thank, perhaps most of all, my mother, Barbara June Kellow Cornell.

PREFACE

> *Death is only*
> *The soul in its row boat, pulling away*
> *Wide water,*
> *Wind glances at its hair,*
> *It is taking*
> *Great gulps of the sky*
> *Embarked it whispers*
> *To itself. Embarked*

—Only Death, *Alicia Ostriker*

*A*fter a long illness, which offered no possibility of recovery other than slow degeneration, my mother decided to take her own life on August 25, 1998. She was seventy-two years old. In the last ten years of her life, she had endured Parkinson's disease, breast cancer, and a series of lung illnesses. She would not want me to tell you much about her suffering or her physical decay. My mother had read a number of books written by daughters and sons about their mother's death and she hated them all. She made it clear that if I wrote a book that portrayed her with the disrespect she found in those books, she would haunt me for the rest of my life and pull me into an early grave. So no more will be said about my mother's bodily condition. On the day she died, she left me committed to the promise to write a book, dedicated to her, that would bear witness to the dignity of her death and that her bridge class would be able to understand.

My mother thought about her death carefully. She chose a suicide recipe of alcohol and barbiturates that would let her die intact. She and I discussed her decision for over six months and tried to look at it from all points of view. No one could have been more rational and coherent about the reasons for her decisions than she was. She wanted to separate her exercise of the right to die, as she saw it, from any of the myths surrounding women's suicides. As her chosen witness, I can tell you that she achieved her end. She died without doubt, without fear, and without flinching as she followed the recipe with the precision that I had come to expect from her.

But this is not a book about the right to die. There are complex arguments on both sides of this issue.[1] I leave those articles and books to be written by people whose mother did not decide to exercise that right. Nor is this a book that focuses on the actual details of someone who took her own life. The irony is that those details are mundane. There are two aspects to such a process. One involves getting information about how to follow through with the intent: the actual taking of drugs or some other chosen means for putting an end to one's life. The other concerns the emotional reactions of the people involved. Books like that have been written and they tend to get bogged down in descriptions that detract from the sorts of things my mother wanted me to present to you. She wanted me to witness to the process in which she claimed her own person through an exercise of the right to die. Bearing witness to my mother's death as she saw it—as the exercise of her moral freedom—can be done only indirectly by discussing how one witnesses to the dignity of the other.

And so this is a book about mothers and daughters and intergenerational friendship and love between women. As you will see from some of the stories I use in this book, by *intergenerational* I do not simply mean relationships between living women. My understanding of its meaning is bound up with a common thread that weaves through all the stories, the theme that holds them all together: women's respect for each other's dignity. I want to define dignity very broadly here, because it is a word not only used by Western philosophical traditions[2] but also by many religious traditions. In Zora Neale Hurston's extraordinary book *Their Eyes Were Watching God*, Nanny, one of Hurston's main characters, says:

> Ah was born back due in slavery so it wasn't for me to fulfill my dreams of whut a woman oughta be and to do. Dat's one of de hold-backs of slavery. But nothing can't stop you from wishin'. You can't beat nobody down so low till you can rob 'em of they will. Ah didn't want to be used for a work-ox and a brood-sow and Ah didn't want mah daughter used dat way neither.[3]

Dignity lies exactly where Nanny locates it: "You can't beat nobody down so low till you can rob 'em of they will." Our dignity and the demand for its respect stem from actual resistance, but also from the broken dreams that are turned into hopes for our daughters. Dignity inheres in evaluations we all have to make of our lives, the ethical decisions we consciously confront, and even the ones we ignore. Dignity lies in our struggle to remain true to a moral vision, and even in our wavering from it, since we are still the ones making the evaluation. Nanny's point is that you cannot lose your dignity even if it is, in fact, brutally violated as in the horrifying case of slavery. Because it is part of the idea of humanity, dignity is something that we cannot lose.[4]

But so often, in relationships between women, and particularly in inter-generational friendships and loves between women, we fail to respect one another. Often in the name of protecting younger women from their "sex,"[5] or in giving advice about how to survive, women betray the dignity of their daughters and those who come after them. As the poet Alicia Ostriker writes:

I write in rage against my sex—
What else can I do? Friend, if your frightened woman
Won't come across, because she's pretty strong, because
She has the whip hand over you, because her mother
And her mother's mother told her: *Use this power,*
Honey, it's all you got, you better make
Them somersault and juggle, make them beg, calculate
Every move or they'll brutalize you, utilize you, take
What you can, sweetheart, give nothing you don't have to—[6]

And then, the next generation, the daughters, both symbolic and real, rage back. One of the central hopes of this book is that we can find a way beyond the repetition of intergenerational rage and resentment by bringing dignity home into the most intimate relationships between us. But if it is important to bring dignity home, it is also crucial to remain faithful to what it demands of us in our attempts at multicultural, intracultural, and transnational dialogues between women. A similar dynamic, but now one of righteous rage, develops between white Anglo women and women of color when the latter accuse us of being arrogant in such a way that refuses them dignity. Often, white Anglo women respond to this accusation of racism with resentment. But the intention to help another woman, if it neglects her dignity, wrongs her in a way that the excuse of help cannot overcome. So if I wish to bring dignity home, I also wish to introduce it as one of the necessary conditions of any kind of movement or reform of women's conditions, not only in the United States but in the world.

Another theme runs throughout the chapters of this book: the integral connection between dignity and mourning. We cannot mourn those whom we do not respect. Of course, we can grieve for them, but mourning, at least as I am defining it, demands that we recognize that there was someone else, someone other than our fantasies of them, that we have lost.[7] Often when we look back through the history of women's lives we seem to find a grim wasteland of broken spirits, victims of their own internalized oppression. But when we impute dignity to those souls, our vision of them changes. Their worth appears to us in such a way that we can, at least, undertake to excavate, or when that fails, imagine, who or what they might have been in their struggle. If we are to remember, we must learn to mourn. Yes, there have been many women, in the words of Nanny, who were so constrained by circumstance they could not begin to fulfill their dreams of womanhood. Actual slavery is an extreme example, but we all know, from the history of our own mothers, the *should have beens*, the *could have beens*, of an unrealized life. My mother certainly had her own. We all do, but that does not undermine our dignity. We can mourn for what might have been within the limits imposed by respect.

It saddened me to see how my mother was so psychically constrained by the bonds of femininity that she had no space to dream. The moral and psychic space to dream and make sense of one's own person—which I was later to name the "imaginary domain"[8]—was closed down to her. By the imaginary domain, I mean the moral and psychic space to reimagine and express the person whom we seek to be. She actually liked the phrase as a description of what she had lost. Perhaps my mother's loss of her chance to dream is what unconsciously drove me to name it, so that we could understand her loss together. Sometimes, in our last discussions, I raved at the limits imposed on her life. My mother was a powerful, determined, brilliant woman and she could have done anything. Whenever I went on a rampage of condemnations of a world that had been so unjust to women, she would simply say, "I have had my life." During one of my most sorrowful moments, as I prepared for her last days, she tried to comfort me by saying, "You are mourning for my life, not for my death." She was right. Sometimes I am asked why I am a feminist. For me, feminism is something I inherited from my grandmother, something I bring to my mother as a tribute, and that which I owe my daughter so that those who may come after her will be able to dream what we cannot yet dream and live out that which we can barely conceive.

My mother gave me a gift in her demand that I write this book. I did not know that at the time I promised to write it. The gift was that she forced me to write in accordance with the respect that the dignity of her death demanded. Her greatest legacy was to teach me to remember throughout this book, and in my own relationship to my daughter, that respect for the dignity of other women already begins to change the way we imagine, live out, and mourn our relationships with each other. I thank her and I remember.

Chapter I

THE INHERITANCE OF DREAMS

In the catacomb of my mind
Where the dead endure—a kingdom
I conjure by love to rise

—*Samuel Menashe*

INTRODUCTION

$\mathcal{W}$ hen I was a little girl, I spent part of every summer with my grandmother. She often took me with her to work and I remember loving our bus rides together. My grandmother was not much of a talker, or much of a moralist, but she constantly repeated one lesson to me on our way home: "Every woman has dignity, no matter what her station in life. Just because she has fallen on hard times does not mean she has lost her dignity. Women want their dignity recognized; they do not want pity."

My grandmother considered the unheeded dignity of a woman a sin; no worse harm could come to her. At the end of her life, she finally called herself a feminist, but was critical of the kind of feminism that pitied women, even if in an effort to protect them from harm. She was, for instance, adamantly opposed to the feminism that portrayed prostitutes and porn workers as women without dignity who could therefore be saved from themselves.[1]

I inherited that same conviction from her: for me, too, feminism must begin with the recognition of the dignity of women. At first glance, this seems to be an obvious point. The word *dignity* and the demand for that respect often appear in feminist literature. What feminist would disagree? But it is not so easy to be true to my grandmother's lesson. Over and over again, we see feminists denying the dignity of the very women they seek to help. The feminist debate over the plight of women sweatshop workers in the free trade zones is an obvious example. A program that recognizes the dignity of these women and their right to self-representation will look very different from one that sees them as pawns of globalization in need of rescue by enlightened Western feminist organizers. I will devote several chapters to thinking about how dignity must inform the ways in which we organize for social change. Indeed, this book is intended to drive home the practical significance of this lesson. Over and over again, I will strike the following note: the recognition of the dignity of other women is an ethical limit to feminist politics. Of course, the question arises: What do I mean by *dignity? Dignity* is an old-fashioned word, used in some religious and philosophical traditions to respect or revere our uniqueness as human beings who can take rational responsibility for their lives or have an inner divinity that gives them infinite worth.

Zora Neale Hurston's character, Nanny, in *Their Eyes Were Watching God*, tells us that slavery prevented her from realizing her dreams of womanhood. But slavery did not keep her brain from raveling ways to undo the horrific injustice of enslavement; slavery did not keep her from shining on in her glory.[2] She mourns the opportunities in life that were so brutally taken away from her. In her mourning, in that alone, we confront her dignity, which cannot be extinguished. Her mourning is a wholehearted condemnation of the appalling graphic effort of slavery to crush her burning assertion that there must be a world worthy of her dignity. As it is a judgment on the world, and an insistence on her worthiness to be free, her mourning also demands a transformed world. Dignity inheres in our right to live free as beings who inevitably evaluate our world and our place in it, and also mourn, as Nanny does, for what has been denied to us.

Put somewhat differently, dignity is a claim on ourselves that signals a world that might be faithful to our freedom. To claim dignity when it has been denied, as in Nanny's case, is already to transform the world. Once the dream is dreamt, the demand is made; it begins to echo in the ears of others. In slavery, the glory of what could not be crushed was often echoed in songs: "I sing because I am free." And the song itself bursts out with the uncrushed wishes and longings that proclaim the "truth" of the dignity of the singer.

Our dignity always appeals to a normative standard inseparable from the ideal of freedom, which inheres in the idea of a creature whose brain ravels away in judgments; whose heart continues to dream; whose soul mourns loss, no matter how bad things get. The great gospel songs of the civil rights movement exemplify the proclamation that we must overcome injustice in the name of the dignity of each and every human being. Obviously, the world we live in refuses to recognize the dignity of each one of us. Thus, we need to ask ourselves: what are the conditions under which actual human beings are given the chance to affirm their dignity, to present and represent themselves as demanding respect for their dignity, and to adhere to the ethical demand placed on them by the worthiness of others to do likewise?

As we seek to change the world, to end the harm women have to endure under conditions of inequality and the denial of their freedom, the recognition of our dignity should operate as an ethical limit within feminism. It becomes particularly crucial to recognize the dignity of other women as the reality of globalization brings feminism inescapably into the arena of international politics.[3]

As a white Anglo woman who is a citizen of the United States who adopted a daughter from Paraguay, I have been forced to confront the harsh realities of globalization and their meaning for women's lives. Given the brutality that women suffer throughout the world under global capitalism, some feminists, particularly in the human rights movement, have clearly been tempted to "save" women from those harms, regardless of what the women involved have to say. The danger, though, is that despite our best intentions, Western feminists engage in the so-called civilizing mission inseparable from the horrifying history of imperialism.[4] The recognition of other women's dignity is a constant ethical reminder that we can wrong those we seek to keep from harm. By denying them the respect we often presume for ourselves, we wrong them. This responsibility to recognize the dignity of other women is one that all feminists have.

Those of us within the United States have a special ethical task. Our country is all too often part of the problem as it sponsors export free-trade zones in which workers, the majority of whom are women, are exploited.[5] Given the shift in the female labor force that it has brought about,[6] globalization changes the issues for feminists. Immigration to the United States is a feminist issue. It is a feminist issue because so many of those seeking residency in this country are women, and also because the issue of immigration has been brought into our homes as many professional women in the cities turn to immigrant nannies to take care of their houses and children. Just think of the recent case of Linda Chavez, George W. Bush's first choice for labor secretary and, before her, Zoe Baird, Bill Clinton's nominee for attorney general. Both of these women were unable to accept their nominations because, formally or informally, they employed "illegal aliens." When we seek to mobilize around these issues, we have to confront the role our own country has played and strive to change the continuing attempts of the U. S. government to dominate the economies and legal systems of other nations. But we also have to face how we benefit from having the prospect of hiring a woman of color, perhaps an undocumented immigrant who has no other choice but to accept a low-paying job as a nanny or housecleaner in order to survive. We have to admit to our role in the responsibility of perpetuating their lives of hardship, which include separation from their own families. Some members and participants of *Unity*—a cooperative organization of housecleaners whose stories I have included in this book as direct interviews—attest to how difficult the lives of these "other mothers" are. It is mainly white Anglo women who are their employers. Inequality is not only "out there"; it is in our homes.

The intimacy of this inequality may explain why it is so difficult for feminists to be true to the recognition of other women's dignity and to frame our political and legal reforms in accordance with respect for that recognition. In a world in which women suffer so greatly it may seem a luxury to worry about the manner of stopping the harm they endure.[7] The desired result— ending the harm—may seem to override morally any other consideration. But another subtler problem has to do with the way in which we make compromises—both consciously and unconsciously—in order to survive in conditions in which women's dignity is denied, a situation that often makes it difficult for us to affirm and demand respect for ourselves. It is tempting, then, to repress those compromises and, as a result, project onto others our own difficulty in affirming our dignity. We all settle with a world that does not treat women as it should, and it is sometimes difficult to face the fact that even as feminists, we still make those concessions. Sometimes we stay in marriages in which the love has gone, but the financial reality of divorce in a world that continues to discriminate against women makes separation a difficult choice. Sometimes we remain married to a philandering husband who publicly humiliates us. Sometimes we enter sex work as the best-paying job we can get and let an abusive pimp take too much of our earnings. At other times, we allow a woman to be wronged in front of us and let it slide in silence because we do not want to be labeled a "troublemaker." When our boss sexually harasses us, we say nothing for fear of a retaliation that will affect us professionally. We let a racist remark about a colleague go by. We closet ourselves, and our partners, because it is still so difficult to succeed in many professions while living as an out-of-the-closet lesbian.

The lives of my mother and my grandmother taught me the need to recognize the dignity of other women. My grandmother understood—this was her lesson to me—that in conditions of inequality and denial of our freedom, the compromises we make should not add up to the loss of our dignity. They cannot justify wronging a woman by denying her dignity in order to "save her from herself." In the case of my mother, the means she chose for her death put me through the ethical test of whether I could stand by my own demand to respect the dignity of another woman. For a long while, I have struggled to come to terms with my relationships with both my mother and grandmother, as well as with their relationship to each other. I have learned how difficult it can be to respect the dignity of another woman and how ethically necessary it is, if we are to break out of the circle of disrespect that denies women their freedom. Of course, to write about my mother and grandmother and the moral lesson of dignity they passed down to me is already to intervene in the debate over whether so-called public values, such as freedom and dignity, should be part of family life and culturally, socially, and legally respected as such.

THE RITE OF REMEMBRANCE

Daring to Do

Because she was a widow, and a wealthy one at that, my grandmother stood in a place in life that enabled her to seize the chances that were usually denied to most middle-class women of her generation. For centuries in the West, widowhood had been a position from which women could exercise power and influence usually denied to single or married women.[8] Not all widows, though, would have had my grandmother's courage or business know-how. The daughter of German immigrants, she grew up in a German working-class neighborhood in St. Louis. Her mother, Amelia Coughlin—nicknamed Mamie—was a domestic in private homes until she got a job as a hotel maid. Her father worked on the railroad. Perhaps because of her background, she knew how rarely life offered chances and she knew how to take one when she saw it.

My grandmother was a latchkey kid. From the age of eight, she had to pick up her siblings from school, shop, cook dinner, tend to her younger sisters and brothers, and take care of her ailing older sister. Her sister's health collapsed when my grandmother was fourteen. The doctors believed her older sister, Margaret, had tuberculosis. She had become so ill that she needed full-time care from her mother. The family's survival depended on the income of both parents. This forced my grandmother to drop out of high school and take a job in a factory in order to alleviate the family's financial crisis. She had to replace her mother's salary, which came from her job at the hotel. She was brokenhearted, not only because she had to leave school, but also because she had to abandon her dream of becoming a mathematician. Still, she did not give up; she worked in a doll factory during the day and pursued her high school degree at night. In her twenties, she found work in a book bindery and took business classes at night. If she couldn't become a mathematician, she could at least move on from factory work to becoming an accountant. Eventually Nana, as I called her, would convince the family to move to California because of her sister's poor health. My grandmother saw California as the land of the future.

Once in California, she went to work at Kellow Brown, a printing company owned by my grandfather. He came from a family of Scots-Irish religious dissenters. His mother, Jemina Kellow, was a well-known Christian Science healer. A believer in the Progressivism of the 1920s, he treated his workers much better than the bosses of the other factories where my grandmother had worked. Accessible to his workers, he ate lunch every day in the same restaurant where they ate. Like my grandmother, he was from a poor family and, like her, had worked at many different jobs before "making it" and ultimately becoming president of his own company. My grandmother told me he loved acting and kept it up as a hobby, dressing up and horsing around during holidays and family outings. For several years, he was a vaudeville actor. I do not know how my grandfather went from being a vaudeville actor to the extremely successful businessman he had become by the time my grandmother met him.

In her usual sketchy manner, my grandmother told me how she applied for a company office job that had been advertised. She spent an entire week's wages on the fanciest dress she could afford—an "investment," she called it— and approached him at the restaurant. She got the job, and six weeks later they married. He proposed to her on their first date, which they spent riding on the recently built roller coaster at the then new amusement park in Venice, California. He picked her up in one of his vaudeville outfits, which included a red coat, baggy clown pants, and oversized clown shoes. Warren Kellow had already been divorced from his first wife, Gertrude, for six years before he met Nana. He had a young daughter, Ruth Gertrude, who was very ill with tuberculosis. Her health began to seriously deteriorate the year after my grandparents' marriage. That same year my grandmother had her first miscarriage.

In 1923, my grandparents and Gertrude Kellow went to Switzerland for almost a year, seeking to do everything possible to save their young daughter's life. My grandmother was not a writer, but she left me a short journal that she kept during their stay in Switzerland. From the journal, it is evident that the two women had an amicable relationship that grew into a friendship as the three tried every means available to them, both spiritual and medical, to save Ruth's life. Ruth called my grandmother Aunt Midge and clearly related to her as a mother figure. In one of the most moving entries of my Nana's diary, she wrote:

> Gertrude finally fell asleep today. I went in to feed Ruth so both Warren and Gertrude could get some sleep. She kept asking me, "Aunt Midge, am I dying?" I kept telling her, "No, your dad can do anything," but she was coughing so hard that the blood kept getting on her food, on her face, on the napkin. I tried my best to hide it from her. Why does someone so young have to die in this horrible, horrible way? Warren won't give up hope, and I try to hope for him. But that cough, that dreadful cough.[9]

At the end of 1923, the three left Switzerland, apparently to take Ruth home to die. Later that same year, my grandmother had her second miscarriage. She was haunted by that year in Switzerland. I knew Ruth died when she was fourteen. I also knew that Gertrude died soon after. But my grandmother would get only so far into the story; she was unable to continue. As tears ran down her face, she would say, "It's so sad, it's so sad." Then, as if to save herself from grief she could not bear, she changed the subject. Only many years after her death, after a close examination of graveyard records during a visit to the family graveyard, did I begin to have an idea of what happened. Ruth Gertrude died on October 31, 1924. Her mother died on November 2, the same day Ruth was buried. It seems that Gertrude killed herself over the unbearable grief from her daughter's death. Both deaths clearly grabbed at my grandmother's heart with such wrenching pain that she simply could not tell me the whole story. On November 27, 1925, after a thirty-seven-hour labor, during which my grandfather was almost arrested for bursting into the labor room (at that time men were not allowed in), my mother, Barbara June Kellow, was born. Until my mother's birth, my grandmother continued to work at Kellow Brown as the accounting vice president.

Warren Kellow was so thrilled at having another daughter that he always wanted to make sure either parent was always with her. He loved to play golf, and long before there were such things for purchase, he fashioned a carrier that allowed him to strap my mother onto his chest so that he could take my mother along while he played golf. After all they had gone through, it was not surprising to hear my grandmother describe the day of my mother's birth as the happiest day of their marriage.

Nana's family did not accept Warren Kellow easily. He was a divorced man, and Nana's devout Catholic sisters were horrified when they heard she had married him at City Hall with only her parents and his mother in attendance. She told me that breaking with the authority of the Church had not bothered her because she actually considered the priests "hypocrites." According to Nana, their "sin" was that they did not respect the dignity of women. Since they did not respect the dignity of women, she refused to take their condemnation to heart. But Warren Kellow changed the material circumstances of Nana's family so drastically that the cloud over their marriage soon dissolved.

My grandmother always told me that women were entitled to pursue their own goals and that she had three main goals in her life. She called them goals and not "dreams," as her family defined them, when she first spoke of them as a little girl. She wanted to bring her family out of the working class. She wanted to own her own business. And she wanted to see every country in the world. A subsidiary goal was to live by the water.

Her marriage indeed brought her family out of the working class. Warren Kellow bought her mother the dress shop she had always wanted, and she left her maid's job at the hotel. Several other family members went to work in management and sales at Kellow Brown. Her father was able to retire. Nana was pleased to boast that she had missed only about fifteen countries in her world tour, depending on "how you counted the countries and whether or not you called them nations." On their tenth anniversary, my grandfather gave her a beach house. They both traveled extensively together. I once asked my grandmother why Warren Kellow and Gertrude Kellow had divorced. She thought about it for a moment and said seriously, "Well, he and Gertrude were different. Gertrude did not like to do things like going up in a hot air balloon." I must have looked so shocked that my grandmother smiled and continued, "Didn't I ever tell you how Warren and I used to travel up the coast of California in hot air balloons?"

From what I can tell, they were well matched. My grandmother was always up for the next adventure. They were both from poor families and enjoyed the freedom money gave them. When I used to press my grandmother about why she loved him, she would laugh as if remembering: "He was always full of surprises." And why did he love her? "Because I was the only one who could always recognize him no matter how he dressed up." Their last five years together were seemingly happy, with that precarious kind of happiness that comes from the shared survival and overcoming of grief.

Soon after their tenth anniversary, and after a game of golf, my grandfather died suddenly from a massive heart attack. It was 1931 and the country was in the throes of the Depression. At the meeting of the managers of his estate, my grandmother announced her intention to take over Kellow Brown as its president. Outraged at her intention, they rose, shouting. Recognizing that in her position as the only woman in the room she could not quiet them down and convince them to listen, she walked out of the room. The following month she took over the presidency of the company. Her mother sold the dress shop and moved in with her to help raise my mother. My grandmother never gave me much advice. But she did pass on a bit of wisdom from that particular meeting: never let men tell you what is possible; men do not have a license on possibility. She was a free spirit when few women dared to be.

My Grandmother's Eva Perón

My grandmother's spiritedness came out in her undertaking to meet and support Eva Perón in Argentina. Although she did not care much for Juan Perón himself, she saw Peronism as a powerful people's movement and became intrigued by the persona of Eva Perón. As a result, she hired a translator and joined the crowds outside the Casa Rosada to hear Eva Perón speak from the balcony. She was mesmerized by her eloquence and by the way she seemed to speak directly to each and every person in the plaza.

After hearing her speak, Mildred, my grandmother, became determined to meet her. My grandmother bought her way into a dinner to honor the Peróns. She actually marched past the security guards, waving them away. Although she spoke no Spanish, she somehow managed to express her admiration for Eva Perón as someone who respected the dignity of the working-class people. Eva Perón invited her to sit down and they spoke long enough for Eva Perón to tell her about her charity projects and other more specific ways in which she was trying to help the people of Argentina. Several days later, my grandmother joined her on a train ride to the countryside, where Eva Perón was to preside over an event in which running water and electricity would be installed and turned on. Along the way, the train stopped and Eva Perón threw money out the window. My grandmother stood up and joined her.

Eva Perón often sat at a table in a public square where people lined up to tell her their dreams; if she could, she would help them realize these hopes. This is how Eva Perón became popularly known as the "dispenser of dreams." On several occasions, my grandmother sat beside her in the public square and joined her in the project of dispensing money for medical care, college educations, wedding dresses, and the like. I grew up hearing stories about people walking away from her in gratitude and hope, with tears streaming down their faces. Over and over again, I heard about the celebrations in the streets as the lights and water were turned on for the first time. My grandmother emphasized what a miracle it was for the people to have these basic utilities. They had never had them before.

Aside from her contributions to Eva Perón's charity, she deeply believed, as Eva herself did, that Eva should become vice-president. When my grandmother heard she was dying of cancer, she flew to Argentina to join the masses of people in their daily vigil outside the *Casa Rosada*. My grandmother's Eva Perón was a working-class girl who took the dignity of ordinary people seriously because she knew what it was like to live without respect. She was a young woman who had accomplished so much in such little time. She was a working-class girl who had not only brought herself out of the working class but who remembered her origins and tried to act consistently with her own memories of how she had been wronged. My grandmother's Eva was also one who fought for the right for women to vote. In a famous photograph, she is shown voting from her hospital bed. My grandmother described how the very ill Eva Perón came to the parliament to watch women be seated in elected positions for the first time in Argentina. Nana, who was sitting several rows behind her, told me how she saw tears stream down Eva's face. Eva was so moved by what the occasion represented for women. My grandmother let no one, including me, speak of Eva Perón's sexuality. She told me that, in her day, there was no way for a woman to ascend to power other than by way of a relationship with a man.

Her identification with Eva Perón was how she integrated and passed on to me her vision of feminine sexual difference and power. Both my grandmother and Eva Perón were clearly ambitious women who wanted power. My grandmother's silences were due to the fact that no woman in her generation could acceptably articulate such desire for power. Given the taboos on the subject, my grandmother taught me both that feminine sexual difference and power could go together, and that a woman who put them together was also putting herself at risk. She did not teach me this lesson by reference to herself, but through the story of Eva Perón.

She saw cynical interpretations of Eva Perón as a sign of class privilege and a denial of the difficulty of escaping the limits of class, particularly if you were a woman. The class identification my grandmother felt with Eva Perón eclipsed the fundamental differences in their nationalities. She explicitly recognized that she was in Eva's territory; she helped her on Eva's own terms. For my grandmother, every gesture she made in support of Eva Perón's programs was meant as a sign of respect for her. In my grandmother's mind, she was not acting as a U.S. citizen helping out a charity in an "underdeveloped" country. She was, instead, promoting the empowerment of a woman *within her own context*. Eva Perón saw fit to include her and recognize her as an ally. As my grandmother told me, using Eva as an example, "Women claim their dignity despite all the obstacles." If it seems difficult for the contemporary reader to give Eva Perón her dignity because of all the fantasies that circulate around her life and her dead body, then perhaps that difficulty must become part of the story.

My Grandmother at Work

My grandmother loved printing and she loved running Kellow Brown. She took me to work with her on a regular basis. I adored standing side by side with her and watching the presses run. At business meetings, I sat in the back of the room and read books. I grew up with the vivid image of a woman exercising authority over men on a daily basis. My own father worked for her for a good part of his life. Her love for the company was "abnormal," or so the rumor went; people thought she should step down and turn the business over to her son-in-law. Such a thought never crossed her mind. When she died, she was still "chairman" of the company. Edward Bergstrom, a longtime family friend and employee of Kellow Brown who had been with the company since 1972, fondly recalls the eighty-something Mildred Kellow coming into the office at least twice a week to check the books. He still remembers her grace and kindness at board meetings. He stressed that whenever she came into the company, she would circulate with all the workers, making sure to have an encouraging word for everyone. For lunch, she always ordered a rare steak sandwich on buttered white toast. She liked to recall with Ed that this was the restaurant where she had first met and addressed Warren Kellow. Ed described the restaurant as a "greasy spoon; it couldn't be any greasier." Once she got her sandwich, Ed remembered that she would then join the workers. Sometimes she would eat with the secretaries or the other office workers; sometimes she would go upstairs to eat with the production workers. Ed told me how she was not only admired but also deeply loved by all of them, "because she sincerely respected all of the workers."[10]

I grew up taking for granted that a woman could be head of a household as well as president of a company and, perhaps more importantly, that she could do it in her own way and in accordance with her own values. Only as society tried to pull me into the strictures of femininity did I realize how much my grandmother had bucked. Few women in my generation have had access to this sort of vision of their mothers and grandmothers. The images can never be dissolved. They challenged the messages and other signals I was later given about the limits of a woman's life. Of course, I heard them. But I always saw other possibilities.

Nana took her family out of the working class, but her class background continued to shape her behavior as a boss. Because she believed her workers needed benefits, she recognized the union without resistance—an action that enraged other members of the printers' association. She also maintained that firing a worker was execution at the workplace; no infraction could deserve such retaliation. She never fired a single worker in the fifty-odd years she ran the company. Her husband had bought houses, which he rented to the workers of his company at below market value. When she died in 1984, it was discovered that she had never raised the rent on many of them. If someone was unable to pay, she let the person live there for free. Some people had not paid rent for years. The market clearly did not govern all the decisions my grandmother made. Dignity and the need for its respect were the ethical mandates that governed the way she ran the company and the properties she owned. Yet despite her unorthodox approach to the demands of capitalism, the company thrived. My grandmother attributed her success to the respect she gave to all those who worked in the company.

My grandmother's egalitarianism was in her guts, and her respect for the dignity of all people overrode any racist opinions she might offer. Despite the Los Angeles union's opposition, which lasted throughout the early sixties, she hired African Americans in skilled production jobs when no other companies and employers did. When the African American elevator operator in her office building mentioned his dream of sending his two sons to college, she offered him financial support. I learned of this only when his sons, both college graduates, thanked me at her funeral. She had simply given them the money, with no thought of being paid back. I have no idea how many other people my grandmother helped financially; never talking about such things was an essential part of her ethos. But I also grew up with the sense that she did not have the words to talk about some of her experiences.

My Grandmother's Silences

Now often referred to as the second wave, the feminist movement of my generation notably emphasized the ways in which women have been silenced.[11] Women's experiences are either lost altogether or distorted by language that is inadequate to the way women seek to represent themselves. This claim has often been taken literally to mean that women do not speak, an assertion that leads to puzzlement or often to a series of examples of eloquent women who speak out in business, in the academy, in law, and in medicine. My grandmother is an obvious case in point. She did not seem silenced by the conventions of a woman's proper role established in her day. And yet I felt that around certain issues, she fell into a silence that could not be broken, even by my continuous questioning and my profound desire to know who she was. The answer to the puzzlement, then, is not simply that women's voices— greater in number and diversity, and addressing various aspects of our social and public lives—are a recent phenomenon.[12] Although this is clearly part of the answer, the meaning of silence is neither so literal nor so shallow. The words we need to say "it"[13] are not readily available to us. We either do not know that there is something there for us to talk about, or we feel our experience recedes before the lack of language.

For most of her days, my grandmother did not allow herself to know how daring her life was. "Allow" was her word. She would have been much too scared to admit exactly what she was doing. But *what exactly* was she doing that required *daring?* She chose to be literal, to describe herself modestly as a woman supporting her family after her husband's death. But the investments she inherited from him had made her a wealthy woman. She clearly did not need to work. As a child, I was puzzled: Why did she pretend she had to work as a matter of necessity? I now understand that in this way, she hoped to bring less attention to her boldness. She had dared to defy what she saw as taboos, things best not talked about. Her modesty was not a completely deliberate guise, though a guise it was.

Still, there were also literal silences between my mother and my grand-mother: about sex and marriage, and about how my grandmother had found the courage to run her own business. But the two kinds of silences—that which is known but not spoken and that which is inaccessible to conscious-ness altogether—are not easily separated. Because her life surpassed what she could say about them, my grandmother did not know how to talk about these subjects. These silences took their toll on the relationship between my mother and grandmother. My mother thought Nana had taken something away from her; something she had that my mother was not allowed to have. What my grandmother had or did not have was precisely what none of us could pinpoint. I wanted it too, although I did not know exactly what it was. There was a gap in what they could say to each other. I inherited that gap.

I was the one who first pressed my grandmother to talk about her life. Until the movement I became part of brought meaning to the word for her, she never thought of the word *feminist*. Once she thought about it, however, she took it on as an adequate description of how she saw the world and tried to live her life in it. It was not that Nana did not want to teach my mother about feminism or help her claim the freedom she knew had been so impor-tant for herself. Rather, it was that this kind of self-knowledge required words that were unavailable to her. There were gaps in my grandmother's self-knowledge that she sensed but could not fill in. She had nightmares about my mother being taken away from her. Feminism enabled my grand-mother to see that she had lived beyond[14] the norms of femininity,[15] in part by never letting herself fully realize what she was doing. And she now had to face the toll on my mother's life exacted by her willingness to let her daugh-ter follow the path laid out for her by those very norms. My mother was only too clear in her later years about how her own conformity had confined her.

The Fifties' Script of Femininity

For most of her life, my mother played by the contemporary rules of femi-ninity because she could not imagine another way to live. They hardly seemed like rules. In fact, they did not rise to consciousness until they were challenged by feminism.[16] Women internalized them as an inevitable way for women to be in the world. My mother came of age during World War II and its aftermath, when the lessons of what is now known as the first wave of fem-inism had been lost to popular consciousness.[17] The women who had worked in the factories were forced to leave their wartime jobs to make way for the homecoming soldiers. They were not just sent home; they were sent home with instructions concerning the rigid boundaries of femininity, concerning the norms that prescribed what it meant to be a man and what it meant to be a woman.[18] The so-called natural differences between the sexes explained, quite simply, why their life prospects diverged so sharply. Nature took the guise of moralism. What was unnatural for a woman to do was conflated with what was abnormal or downright immoral. During that time, a "good woman"—heterosexual by definition—was to live within the appropriate limits, smiling all the while.[19] These norms regulated a woman's sexuality and shaped her life as wife and mother. They had little to do with the quest for moral choices or with the search for the good life. Indeed, the very idea of that quest was itself a challenge to the conventions of a woman's place. To undertake such a journey was to demand a level of individuality deemed inappropriate for the middle-class white woman—idealized as she was in the United States of the fifties.

These moralistic prototypes of womanhood were never easy to live up to, and they were often contradictory.[20] Dual roles as good mother and sexy wife seemed in conflict. The guidelines were for middle-class white woman only, assuming they could ever hope to live up to them. Failure was built in. Nevertheless, these norms for the "good woman" were enforced at every turn through kinship structures, schooling, laws, economic strictures, and the cultural weapons of television and women's magazines.

If the ideals of femininity were complex and mysterious, the bottom line was that women were destined to be wives and mothers. My mother married at eighteen; my father was going to war, and he desperately wanted to marry her. She had started college that fall and had earned straight A's her first semester. She also rushed for a sorority, an experience she remembered with loathing for the rest of her life. In the course of running for the right sorority, women were expected to show that they were perfect young ladies. You had to wear the right dress, the right jewelry. Your hair had to be just so. You had to nod in a certain manner, and on top of all that, you had to be able to walk daintily. My mother's head spun; she feared she would never get it right. If she had the right dress, most likely she would be wearing the wrong shoes. And then, there were always those girls who had already made it into the sorority and were judging her. To complicate matters, her name was Barbara, like so many white women of her generation. To make sure I would never blend in and get lost in the crowd during rush, she named me Drucilla. I never did rush for a sorority, but the name has served me well.

Escape From Pressure

My mother got pneumonia twice while she was rushing. The two spells of pneumonia made Nana worry terribly about her daughter's health. My mother had long suffered from lung ailments. She had first come down with a mysterious illness soon after her father died. At the beginning, it seems, she was prone to severe asthma attacks. Since she was a Christian Scientist, my grandmother first took her to spiritual healers. But not being a strict adherent to the religion, she also took her to medical doctors. They kept changing the diagnosis. At one point, she was told that my mother had a rare form of tuberculosis. Given her past experience with the sickness, the diagnosis made my grandmother desperate. The doctor concluded that my mother did not have tuberculosis, but remained unclear about what she had. Many years later, my mother was diagnosed with bronchiectasis. By the time she was correctly diagnosed, she already had a colonized infection. It was the inability to treat a colonized infection that led her to a slow and steady deterioration. If bronchiectasis is diagnosed early enough, a colonized infection can be avoided.

My mother's illness was ultimately attributed to some kind of feminine disorder. Nana feared that some day the doctors would find out what my mother really had, and that when they did, it would be dreadful news. She was so anxious that she hovered over my mother. Apparently, my grandmother was an overly protective and domineering mother: fear of the horrible disease and of death was undoubtedly part of the reason. I know the two bouts of pneumonia my mother suffered had scared my grandmother out of her wits. This probably explains her willingness—so out of character for her—to let my mother drop out of college. It was not like her to accept such a decision because Nana was firmly committed to the pursuit of higher education, not only for her own daughter and for the sons and daughters of her brothers and sisters but for her employees, many of whom did not even seek out her help. And yet, she willingly agreed to my mother not finishing school.

Nana's Different Personae

My grandmother also tried to determine the course of my mother's social relationships, particularly those with men. During her last year of high school and that first semester of college, a slightly older man, John Church, wanted to marry my mother. At the same time, Clark Cornell, the man who would ultimately become my father, pursued my mother and asked for her hand in marriage. Taken by her intelligence, John Church wanted her to finish college and forget about the sorority rush that was so upsetting to her. But "good girls" were not supposed to forget about sororities and her mother did not approve of him anyway. John Church, who was very much in love with my mother, describes his first meeting with my grandmother:

> However, when I dropped over to visit her at home for the first time, I sensed that there were major obstacles to what I had assumed was a ripening relationship. Chief amongst these was her mother. Mrs. Kellow answered the door to see this grinning newcomer who asked to see her daughter. It was immediately evident that Barbara June had not inherited her empathy and receptivity from her mother. After a cool survey of my appearance and the 1932 Ford at the curb, she asked me to identify myself. After what seemed a long decision-making process, she asked me to enter. When BJ appeared, the three of us went into the spacious living room. A stilted conversation of about fifteen minutes followed. BJ was her usual friendly self, but Mrs. Kellow assumed the role of interrogator.
>
> Her questions reflected little interest in me and seemed aimed at bringing Clark Cornell into the following exchange: "Have you met Clark?"— "No." "Has BJ told you that they expect to be engaged soon?"—"Not exactly." "Did you know he has just received his pilot's wings in the Air Corps?"—"Yes, BJ told me." "Don't you think the country is lucky to have young men of his caliber in the service?"—"Oh, yes." It seemed clear to me as I drove away from the brief call that Mrs. Kellow had her mind set on Cornell as a son-in-law. One of BJ's close friends told me soon after that Cornell had worked summers at the Kellow's company. She said Mrs. Kellow was grooming him to take over the family business after the war.[21]

It became evident that Nana never had any intention of grooming anyone to take over her company. But John Church was right about one thing: my grandmother had decided that my mother should marry Clark Cornell and that John Church should be pushed out of the picture. My mother fought back a little, and even that was rare for her. My grandmother condemned John Church as a "Russian Jew"—he was neither—and my mother conceded that she would not see him anymore. Why did my grandmother choose Clark? Why would it be her decision and not my mother's? Where was the grandmother I knew in John Church's description of the domineering mother who ran her daughter's suitor off?

Obviously, John Church saw a side of Nana that was never presented to me. The Nana I have described so far is "my Nana," the Nana who dwells in my memory, who always supported my hopes, and who never doubted my dreams, even though they often got me into trouble with the powers that be. She would go to any length to support me and was willing to ignore the others—even to the extent of bucking authority—if that was what it took.

When I was nearly four, I decided to marry a tornado named John. John was a bit demanding in that he insisted I always wear a wedding dress. I designed the dress myself. My grandmother took my design to a dressmaker and had twenty dresses made with different colored sashes. Although it seems rather extreme, I needed twenty because I could not wear any other kind of dress throughout the summer. Things went fine through the summer. I spent most of it with Nana at her beach house, and she had no problem with my needing to wear a wedding dress every day. But then, in the fall, I entered nursery school and the nursery school administration had a different view on the matter. They told my mother that I had to wear "normal" clothes, which did not sit well with me since it went against John the Tornado's wishes.

My mother took the side of the school and insisted I put my imaginary world in perspective, something I was not willing to do at the time. The next day, in tears, and after a huge battle with my mother, I arrived at the nursery school in "normal" clothes. Soon after I arrived, I ran away to my grandmother's house, which was very close by. When I got there, I begged her to understand that I simply had to wear my wedding dress. Meanwhile, the school had called my mother to let her know I had disappeared. My mother was quite naturally worried sick about me. She called my grandmother, who reassured her that I was with her and that I was fine. Once my mother recovered from the fright, she was furious with me. My grandmother came up with our own secret compromise to solve the problem. She told my mother that I would go to nursery school in normal clothes. I agreed. However, soon after being dropped off, I would wait behind the stone entrance to the school for my grandmother to come pick me up five minutes or so later. As soon as I got in the car, I would change into my wedding dress and off to Nana's work we would go. Nana withdrew me from that school presumably on the basis of her belief that I did not need to attend nursery school. Only years later did my mother find out about this arrangement. She really thought I was going to nursery school.

This was only one of many secret compromises that my grandmother made over the years in order to protect me from what she apparently saw as my mother's attempt to reshape me so that I could fit into what my mother believed was a "normal" girlhood. Sometimes, though, Nana would not actually act against my mother. For example, my mother felt that my red hair made me stand out too much and had the effect of making me look "odd." And so she took me to the beauty parlor and had blond streaks put in my hair. My mother herself was enslaved by beauty parlors. She was not trying to impose on me something she did not demand of herself. Indeed, she did not like to go on vacations for more than a week because she did not like to have to do her own hair. She was determined that her hair be perfect in the sense that it would conform without flaws to the styles for a proper woman.

Looking back now, there is a question about why my grandmother wanted to compensate for my mother's attempt to "fix" me so I could conform to the world around me, instead of confronting my mother when she disagreed with what she was doing. This question is undoubtedly part of the gaps in the story of why the relationship among the three of us took the shape it did. My grandmother's concern for the protection of my space for imagination and self-expression seemed to operate exactly opposite to the way in which she constrained my mother.

I can only speculate about why this was the case. My mother's lung disease terrified Nana. The irony is that in the name of protection, she smothered her daughter both by indulging and controlling her. I am now convinced that there was an unconscious fear of losing her. She had known death and the sudden way in which a loved one could be taken away. However, I believe there was something else going on, something that took me years to figure out. My grandmother was haunted by the fear that the freedom she claimed for herself could somehow lead her to be punished by society—a society that could take her daughter away from her. Throughout her life, my grandmother had terrible nightmares about my mother being stolen from her and outright killed. Unconsciously, she "knew" that a patriarchal society would not allow her to get away with her "free" life. So she tried to keep quiet about it and made sure my mother remained safely by her side. She saw John Church as an obvious threat because he could not be easily controlled by her and might take her daughter away. Her relationship with my mother was clearly dominated by her fear. Yet she could not name her fear. Unnameable, it haunted their relationship.

When I asked her later, my mother never said she married my father for love, nor did she seem ever to have asked herself the question. My father was a handsome young man, and he was very much in love with my mother. He relentlessly pursued her. His determination, plus my grandmother's approval, seemed to result in my mother's sense of inevitability that he was the one she was to marry. It was this inevitability that she did not question. Even to consider such a matter independently would mean she was making her own crucial choices, a role my mother did not see for herself until the very end of her life. Like so many of the white middle-class women of her generation, she could not even imagine a career appearing on the horizon of her possibilities. Marriage was for her an acceptable way out of a pressure for which she was unprepared.

Yet her own mother not only worked but had been the first woman president of a printing company in the United States, and ran that company for most of her life. Not until much later in their lives did either one of them discuss my grandmother's choice to live the way she did. Fifty years passed before another woman became president of a printing company the size of Kellow Brown. The woman who followed her was my younger sister, Jill Gwaltney, who became president of Forms Engineering in 1983. Forms Engineering was a company initially founded by my father in the late sixties. After she graduated from Stanford University, my sister went to work for him. Soon after she became president, another woman joined her to become one of the two female presidents of a Top 200 printing company in the United States.

Whose Demands?

On the surface of things, my mother and grandmother seemed to be very close. Our family vacations were usually spent with Nana, and as young children we all spent lots of time at her house. According to my brother, Warren Bradford Cornell,[22] we benefited from Nana's love of travel. We all had been taken on fabulous trips to Europe, Africa, and in my case, South America, long before we reached adulthood. After my mother's death, my brother spent months sorting out photos for a CD in order to keep alive the family memories of my mother and grandmother. When he finished, he wrote his observations, one of which was that as her grandchildren, we were beneficiaries of Nana's adventurous spirit, perhaps even at the expense of our own parents' independence. My father worked at Kellow Brown a good part of his adult life and he did not achieve any kind of independence from her until he founded Forms Engineering. With her usual generosity, my grandmother helped my father by lending him what was needed to start his own business. So even in that endeavor, he remained indebted to her.

Until recently, I resented the image others had of my grandmother. I thought that it refused her own daughter the individuality she always claimed for herself. Nana certainly had encouraged that freedom in me. But that was my own experience and now I can see what may have unconsciously motivated my grandmother to hold on to her daughter so tightly. There is no doubt that she was an extremely powerful woman. Had she not been, she would have never accomplished what she did. When she wanted to achieve something, she did not let anyone stand in her way. She was truly formidable in that way. After the death of her husband, she accomplished what she did entirely on her own. Nana loved her family. But since she had grown accustomed to making the decisions for everybody, including her own mother, she did not consider her three sisters and her one brother as confidants, even while confronting the many difficulties facing her.

My grandmother was deeply affected by the loss of her husband. As an adult, my mother often wondered how it would have been had her father lived longer. She remembered that he took her everywhere with him and encouraged her to believe that she could do anything. She was almost certain that he would not have allowed her to quit college at eighteen and get married. I suspect she was right. After his death, something seemed broken in my mother. I do not know if it was ever repaired. Nana never explicitly encouraged or discouraged my mother to follow in her footsteps and enter what she called the "business world." My mother never imagined that her mother's accomplishments were something all women could achieve.

However, my mother did make an early attempt to resist playing the "proper" woman's role. She was reluctant to have children and she certainly did not want three. My grandmother sympathized when it came to numbers. "One, up and run; two, still can do; three there you be," she recited. But she also believed that raising a daughter was the best thing any woman could do.

So it was not from her own mother that she felt the pressure to have more than one child. Rather, it was the result of a general social pressure: a sane woman simply did not declare in public—or in private, for that matter— that she did not want children. When my father, back from the war, was eager to start a family, my mother suggested she might never want children, to which my father countered that she see a psychiatrist. She never went to the psychiatrist, but did what she was told. When she was twenty-one, she gave birth to her first child. Before we got our own lives off the ground, she warned us, her daughters, against having children. She made it clear that becoming a mother was unquestionably a matter of choice.

Even though my mother tried to get me to conform to certain conventional feminine norms, I believe she was less concerned with my femininity than with protecting me from being "weird." She never tried to make my younger sister conform to those norms. Usually, without reference to gender, she stressed excellence for all her children. It was not enough for us simply to try. We were to be on top—period. My mother had my grandmother's drive, but in her case, it was not aimed at her own aspirations but at her children's achievements. Although my mother did not want children in the first place, she invested herself in us completely once we were born. Her last trip in 1994 was made to hear my brother, Warren Bradford, and myself speak at different seminars at the University of Chicago. My brother was speaking in the economics department in a seminar run by the Nobel Prize Laureate Gary Becker; coincidentally I was giving a paper at the law school that same week in their seminar on law and social theory. Indeed, my brother and I did not know that we were both speaking, one on Tuesday and the other on Wednesday, during the first week of November. We told my mother, but she did not interpret our invitations as a coincidence. Instead, she thought of the situation as a tribute to what she had taught us to be. Ill though she was and unable to fly, she was determined to make the trip anyway. Carrying her oxygen with them, my mother and father took a train across the country to Chicago. My mother was in her glory during those three days in Chicago. She told everybody who would listen that two of her children were seen as academically worthy enough to be invited to the University of Chicago. My mother was still well enough to celebrate with me by getting us both a cosmetic makeover at Saks Fifth Avenue in Chicago. It was close to her birthday, so I wanted to buy her all the products they tried on her face. My mother insisted that the "coincidence" of the invitations was present enough. I bought her the makeup anyway. This was our last makeover together.

My mother always fondly remembered the way I looked that day when I spoke at the University of Chicago. On one of the many afternoons we spent talking, during the last month of her life, she interrupted herself and said: "I do know what I would have done differently in my life. I would have been a lawyer." For me at least, she was the most combative person I have ever known. I have no doubt she would have been a great lawyer.

But it was I who went to law school. I had started my life as a writer and had already published a few poems before I graduated from high school. My mother did not think writing was a realistic profession. In the early 1970s, I continued writing, but I was also a student activist at Stanford University. I left Stanford right before my graduation and started working at a factory. Ultimately, I became a union organizer and worked with the unions until 1976. During my time with the United Electrical Workers, we fought a corrupt Teamster local in Hackensack, New Jersey. This Teamster local would offer sweetheart deals with the employers and then promise them to get rid of me. While the workers received no benefits, the corrupt Teamster local received a lot of dues. United Electrical Workers tried to fight the Teamsters every way they could. They were overwhelmed by the lack of resources and the endless threats of violence. My mother was completely baffled by my work with the union and my decision to leave Stanford. In 1976, I moved back to California because I was ill after this long political battle with the Teamsters' union. My mother then saw this interlude of my illness as an opportunity. She suggested that I finish college and go to law school. To her disappointment, I never returned to Stanford. I graduated from Antioch College by correspondence course. I never thought law school was the right decision for me.

The afternoon when she realized what she would have done with her life, we both managed to find the words to talk about how I carried out her unconscious wish for herself. Once we acknowledged this, we were both able to mourn her not having fulfilled her desire to go to law school. In her days, few women went to law school. Even after graduation from college, women were completely excluded from the practice of law. It never occurred to my mother that *she* could become a lawyer. The discussion liberated me from continuing to be an extension of her unconscious wish.

My Mother's Legacy

In some of the most interesting psychoanalytic literature on the subject, the *phantom* is a literal secret that returns to haunt someone in the next generation. Nicolas Abraham and Maria Torok give the example of illegitimacy. The son, who has no conscious knowledge of his father's illegitimacy, becomes preoccupied with his lineage—indeed, with lineage in general. In a famous case, the son of such a father is obsessed with family crests, which represent patrilineal lineage. The analysts put this story together because it was one that could be told. Symbols were readily available to represent a man's place in his line and what it would mean to a man if those symbols of his place in society were missing. The *phantom* is possession by someone else's unconscious.[23]

The *phantom* is almost always a literal secret in the work of Abraham and Torok. The illegitimacy is also literal, a father not claimed by his own father. I have described my grandmother's sense that the exercise of her power as president of her company was something apparently illegitimate. But the symbols to articulate neatly this sense of illegitimacy were not available. In the case of intergenerational haunting between women, the secret may also be what cannot be said. My mother could not speak her desire, so it remained unconscious and I enacted it, even though it was not mine. Once we both realized that I had enacted her unconscious wish, she encouraged me to do what I desired to do. Indeed, a year after her death, I accepted a full professorship in political theory. I returned to writing mainly as a playwright.

My mother was the most competitive person I ever met. She lived long enough to laugh about her competitiveness. As a child, she wanted to turn me into a competitor. When I was thirteen, I threw a tennis match that would have made me the girls' champion at my school and would have entered me in a state-wide competition. It seemed the right thing to do: the girl on the other side of the net was crying and her father was yelling at her for losing. I did not want my own victory to cause such havoc in someone else's life. My mother watched in disbelief. She raged at me only a few times in her life, and this was one of them. She thought that caring for the other girl was something that was getting in my way.

My mother never accepted the idea that women were or should be more caring than men. When a young woman friend of mine in junior high told me that women should never beat men in a competition because men considered such behavior unfeminine, my mother declared that the stupidest thing she had ever heard. She also dismissed the idea that women should mask their intelligence to please men.

And yet, my mother never owned her intelligence. As she grew older, she increasingly saw how her own life had been scripted. She described herself as acting like an automaton, as someone without dreams. Later in her life, she corrected herself: she had not been allowed the space to dream. For years, I struggled to articulate the space that my mother had been denied and that I wanted to ensure for my daughter. I have called that space the "imaginary domain,"[24] the moral and psychic space we all need to become whom we seek to be.

My mother experienced the space of dreams not as closed, but as foreclosed, something so lost it could not even be sought after. She experienced this until the end of her life. Then she began to mourn its loss and to see that her life as a "proper lady" was not one she wanted to continue to enact. Her body was wracked by a twenty-year battle with lung disease; she began to see some of the unconscious, internalized limits that had formed her. She knew she had little time left, but she did not want to have the subjectivity she was just beginning to exercise undermined by a body decaying right before her eyes. Death became her chance to exercise her newly found individuation. Determined to make the decision about her death on her own, she carefully interviewed her doctors about what lay before her. Each had a different definition of what it meant to be terminally ill, so she reached her own. She would die while she still had the strength to take her own life. She got one of her doctors to agree to give her the pills she could use in the recipe she had in mind. Together, my mother and I read a great deal of literature on the right to die. We debated all the issues back and forth. As my mother often pointed out, it was I who sometimes became overwhelmed with emotions and lost the train of my arguments. She told me she never let her emotions get her off track. She had always kept my books on her coffee table, where they seemed to sit forever without being read, but she did read carefully the manuscript that was to become *At the Heart of Freedom*. She used some of the ideas in that book to justify the right to die. I was thrilled and terrified. My mother eventually ended the debate, making her decision the last word.

My mother set a time and place for her death and definitively decided the manner in which she would take her life. She then embraced the death sentence she had set for herself. This acceptance allowed her to break away from the scripts and conventions that had ensnared her. She authored her last days. She told me that now she knew exactly what I meant when I argued that claiming one's own person was a project. For that project, at least, she believed it was not too late.

Even when it ends a terminal illness, suicide is kept quiet, particularly a woman's suicide, bound up as it is with other women's suicides resulting from depression, hysteria, or whatever disorder is attributed as the cause. My mother wanted to separate her suicide from any notion of psychic turmoil and decided to distinguish herself through managing the ritualization of the means of her death. She put it simply: "I will not be taken by death." She resisted the clear societal message that good girls do not commit suicide; as with every other decision in their lives, they are to wait to be taken. But she also wanted to be clear about the fact that she exercised her right to die—that she claimed and owned her freedom. She did not die as a victim of her own illness. It was this exercise of her freedom and *my* respect for her dignity that she wanted me to discuss in this book.

On her last day my mother told me, "I finally become myself in the manner of my death." Soon after, she marched, without flinching, into the abyss. She knew I was to dedicate this book to her, and she wanted me to write about her death. She had few causes, but the right to die became one of them. As she told me, "I did not live my life as a feminist, but I am dying as one." I promised I would write her end as she enacted it, as her own celebration of the dignity she openly and proudly claimed as her own. And so I have, or at least have tried to do so, for she left this daughter with the awe that is integral to the feeling of respect inspired by someone who dies as her own person.[25] The day of her death was the day I learned just how wrenching it can be to persist in the recognition of the dignity of other women, how excruciating that feeling of respect can be.

I have given you the story of this lesson not to violate another woman's dignity, but as it was passed down to me. I have given you a story of gaps and silences, dreams lost and dreams realized, phantoms and secrets. I have undoubtedly been enabled and inspired to write because of that legacy. Of course, my grandmother and my mother's influence goes way beyond the conscious lessons they taught me and the values they passed down. I write to them and through them. Sometimes I have consciously sought words to end the silences I inherited. The gaps in their ability to speak to each other have placed an ethical demand on me to allow them to rest in peace. More than I know, I have probably been working through their history and their relationship to each other. This story begins with the presumption of their dignity and proceeds with the respect that such a presumption demands. Only then can we feel the full weight of what they bore, the encumbrances that made their dignity so hard to claim and that, indeed, could not be claimed without enormous struggle as the basis of every woman's right. Recording my respect for my mother and my grandmother's dignity has compelled me to think about the ethical, moral, and political significance of dignity as an ideal we should all embrace.

Chapter II

DARINGS OF THE FEMININE

In a way, her strangeness, her naïveté, her craving for the other half of her equation was the consequence of an idle imagination. Had she paints, or clay, or knew the discipline of the dance, or strings; had she anything to engage her tremendous curiosity and her gift for metaphor, she might have exchanged the restlessness and preoccupation with whim for an activity that provided her with all she yearned for. And like any artist with no art form, she became dangerous.

—*Toni Morrison,* Sula

$\mathcal{I}$ ntergenerational stories between women are not simply passed down. They are also created and re-created when they are claimed as the legacy of the writer, especially when the writer consciously assumes her inheritance and accepts that it places an ethical demand on her that can be met only in her telling of the story. My legacy is reflected in my struggle to develop the words and ideas to represent the gaps and silences in my mother and grandmother's relationship. As I strove to understand the moral and psychic space that had never been given to my mother, I developed my conception of the *imaginary domain.* Through my mother and grandmother, I learned to live with the emotional impact of the feeling of respect that confronting the dignity of another person demands.[1] They are never *just* with me in the arguments I consciously introduce and defend as reasonable. The story I told in the introduction will go through me and on to my daughter, Sarita Graciela Kellow Cornell. As a Latina adopted from Paraguay, Sarita will bring to it twists and turns that are still unknown. The story will change as she takes responsibility for its perpetuation and as it grows in her, meshing with her understanding of who she is. In her discussion of

"keepers and transmitters," Trinh T. Minh-ha evokes such responsibility between generations:

> Every woman partakes in the chain of guardianship and transmission. . . .
> Tell me and let me tell my hearers what I have heard from you who heard it
> from your mother and your grandmother, so that what is said may be
> guarded and unfailingly transmitted to the women of tomorrow, who will be
> our children and the children of our children. These are the opening lines
> she used to chant before embarking on a story. I owe that to you, her and
> her, who owe it to her, her and her. I memorize, recognize, and name my
> source(s) not to validate my voice through the voice of an authority (for we,
> women, have little authority in the History of Literature, and wise women
> never draw their power from authority), but to evoke her and sing. The
> bond between women and word.[2]

The notion that feminism will lose its way if it cuts itself off from the
intergenerational stories passed down from one woman to another is one
reason for turning to psychoanalysis.[3] At the heart of psychoanalytic theory,
no matter what school,[4] is the insight that human beings grow into themselves *only* through their relationships with primordial others: parents,
grandparents, aunts and uncles, siblings, "actual relatives," or others who
participate in these kinds of close relationships.[5] Before her identification as
a feminist, my grandmother's life as a woman executive and as an adventurer
was inexpressible, even to herself. These gaps in her self-knowledge remained in her unconscious and could not be articulated. In different ways,
my mother and I ran up against those gaps. She seemed to carry a secret.
Clearly, I thought I could not tell their story without reference to what had
remained unspoken between them, in large measure because my grandmother did not have "the words to say it."[6] Obviously, given the richness
and the complexity of human experience, there are many forms in which
laws of separation within families, and rules of differentiation between people through the generations, can be respected.[7] And there are, indeed, many
terms that enunciate and describe the ethical laws used by many divergent
religious traditions between the generations. My own argument is for an extension of the concept of dignity to family and kinship relationships. What
is at stake in this extension is our freedom to keep ourselves from falling
prey to drives that pull us hither and fro, preventing us from collecting ourselves enough to be able to express our desire, let alone pursue it and rationally evaluate it. Addiction to drugs or alcohol is a classic example of being in

the throes of drives that compromise our desire.[8] This understanding of dignity is a reinterpretation of its meaning, which can be made only with the help of psychoanalytic theory and, more especially, the interventions that have been made in that body of theory by feminist analysts.

Psychoanalytic insight defends dignity as the moral mandate by which all of us are viewed as people who *in principle* can articulate their desire, as well as morally evaluate their ends. The articulation of desire has always been assumed as necessary for moral freedom and responsibility. Indeed, much political philosophy takes it for granted that most people act as actively desiring subjects who simply shape their own lives.[9] Of course, some of the earliest critiques of canonical political theory offered by feminists argued that it was easy to make this assumption because the subjects in the purview of the theory were straight white men of a certain class background.[10]

GENDER, SEXUAL DIFFERENCE, AND THE FEMININE WITHIN THE IMAGINARY DOMAIN

Here, then, the question of gender and sexual difference needs to be raised. As we will see, psychoanalysis can help us grapple with the significance of sexual difference. More specifically, psychoanalysis can help us understand why *the feminine within the imaginary domain* can be infinitely represented, and represented so as to explore the culturally and legally imposed norms of femininity. As I have defined it within the legal sphere, the imaginary domain is the moral and psychic right to represent and articulate the meaning of our desire and our sexuality within the ethical framework of respect for the dignity of all others. This domain is imaginary in the sense that it is irreducible to actual space. But it is also imaginary in a psychoanalytic sense. In order to understand the subtler implications of my conception of the imaginary, we need to examine briefly its psychoanalytic roots. Our assumed identities have an imaginary dimension since we envision ourselves through them. Again, to try to keep my terms as clear as possible, I am using the term *imaginary* in the following sense: we all have a self-image that we form through our identifications with others as they have imagined and continue to imagine us.[11] These identifications color the way in which we envision ourselves, but do not determine the reach of our imagination in dreaming up who else we might be. In this way, I distinguish the imaginary from the radical or the productive imagination in which we envision new worlds and configure what has otherwise remained invisible. The radical imagination demands some degree of psychic separation. Otherwise our dreams of who we might become, both individually and collectively, can be captured by unconscious claims on us.[12]

This imaginary is originally shaped through our identification with our primordial others. Let me give an example of an image that inspired my own imagination. It is of my grandmother's authority and her view of me as someone who could assume and exercise it one day. During my childhood, neither my grandmother nor I had the words to explain why that image seemed to run up against a barrier that would not yield—a barrier that did not allow her to articulate how she saw herself and how she was seen. But the images of her walking into her office as "the boss," albeit as a particular kind of boss, remained in my imaginary. Undoubtedly, these images of my imaginary put pressure on me to find the words that would explain how a woman could claim the entitlement to exercise authority as a *woman*. But we can also be hypnotized by "internal tyrants"[13] that bind us to images of ourselves, and imaginary demands on us that can prevent us from having a sense of entitlement and from claiming our own desire. Later in this chapter, I will retell the story of Marie Cardinal, a woman held captive by such an "internal tyrant."[14] We need the imaginary domain so desperately precisely because it provides us with moral and psychic space and, with it, the possibility of both escaping those tyrants and learning how to accept the necessary struggles of our mothers and grandmothers.

Another reason for my usage of a psychoanalytic conception of the *imaginary* is that, at the level of our psychic life, I defend feminism as an ego ideal. We form ego ideals by envisioning ourselves either through real or imagined others. We imagine either that we reach such an ideal or that we can become what the ideal holds out for us as a possibility. Feminism envisions how we might be as free and equal persons in our day-to-day lives. As an ego ideal, it cannot be imposed. Nor can we say that one must act or be a certain way in order to be a feminist. To make such impositions undermines the power of feminism as an ego ideal. To understand feminism psychically is to defend its spirit of generosity because each woman or man will internalize it as an ideal in her or his own way. This generosity of spirit does not directly serve the transnational literacy to which international feminisms must aspire. But the understanding that every woman needs to be respected in her effort to link her feminism with actual attempts to change our world through solidarity does indirectly serve the moral imagination. Thus, we are, once again, returned to the need to respect the dignity of all women as the ultimate ethical law in which feminist political action must proceed in its struggles.

Let me now turn to the feminine within the imaginary domain as I define it. It is that which remains connected yet irreducible to both the requirements of biological reproduction among humans and the cultural laws shaping kinships and family structure. Gender is often used in contemporary feminist theory to distinguish the biological residue of sex or sexual difference from the socially constructed, culturally encoded, legally imposed division of human beings into two kinds, recognizable as men and women. For me, gender is a limited but useful category. We can usefully rely on gender insofar as there are fixed, recognizable identities within societies. For example, we need to know what the wage differentials between men and women are in all aspects of working life in the United States. Gender, then, is a shorthand useful in describing the specific comparative inequality between men and women, which clearly continues to exist in the workplace.[15]

My focus in this book is on the feminine within the imaginary domain and not on gender. I use the phrase "the feminine within the imaginary domain" to focus on the subjective aspect of the assumption of sexual identity, by which I mean the process through which we internalize both an image and a set of norms that shape who we are as well as who we desire and love. This subjective aspect of our identities cannot be easily quantified. In her influential book, *Gender Trouble*, Judith Butler effectively argues that gender identity is never simply and passively internalized as images and static norms.[16] These norms and images are also shaped as we externalize them and actually act out our lives as men and women. Since we not only assume identities but also live them, this process of acting-out is inevitable. We are the ones who externalize the meaning of gender. How we assume these identities is never something "out there" that effectively determines what and who we can be as men and women—gay, lesbian, transsexual, straight, or otherwise. This process of the internalization and externalization is what I mean by the subjective aspect of our lifelong enactments of our sexual difference. The radicalism of my argument follows from the claim that the more we actively assume our desire, the less we are captured by traditional gender roles, and are thus enabled to assume our "special responsibility"[17] for our lives.[18] By desire, I do not mean simply sexual desire, but rather what we broadly conceive as our ability to chart out a life that is our own.

It is also worth examining our understanding of dignity and discussing the debate about the relationship between care and justice.[19] This debate has often been coordinated with the feminist critique of the model of the political subject or person as privileging rationality and independence over connection to and care for others. The framing of this debate may have created divisions that are not as stark as they may seem. The idea of dignity I am defending here functions at a more primordial level. How can it be that there is someone "there" who can *value* care and intimacy over independence and critical reflection? My mother's own account of her history can help me clarify what I mean by *more primordial*. My mother did not experience herself as someone who claimed her own desire until the last months of her life. She had children, not because she valued connection, but because she felt she had to, or face a kind of ostracism she would not have been able to withstand. Simply put, when psychoanalysis speaks of individuation it should not be conflated with individualism.[20] In this sense, my mother did not "individuate" herself and her desire from her own mother's desires and wishes for her.

I will suggest that we replace care, as it has been understood in this debate, with the phrase *ethical and affective attunement* because this phrase captures more precisely the kind of attention children need—care that goes beyond physical care. Parenting requires that we constantly be "tuned into" our children. This is indeed an ethical task that may demand a respectful struggle with "embodied willfulness"[21] that is not easily reconciled with the word *care*. Respect for our dignity and our imaginary domain allows us to individuate enough so that we can claim our desire and take effective responsibility for our lives.

DIGNITY AND DESIRE

In the Grip of the Symptom

I want to offer an account of the suffering of Marie Cardinal, a woman who was unable to connect and express her own desire because she could not individuate herself from her mother. *The Words to Say It*, Marie Cardinal's autobiographical novel of her own psychoanalytic treatment, is very relevant for contemporary feminism. The psychic life of the author is bound up not only with the cultural dressing of femininity but also with the patriarchal culture of French colonialism. Her personal drama, and that of her mother's life, unfolds against the collapse of the mother's profound investment in the "civilizing mission" of French colonials in Algeria. The colonialist background is only sometimes brought to the center of the novel. Still, it has the effect of shattering the mother's fantasy that her life has some meaning because she is "helping" the natives with her charity work. The masquerade of femininity is played out in terms of the social conventions established for an upper class white female under colonialism. This is not, then, just the story of any woman; it is not an account of the meaning of femininity, or of its lack of meaning as a general matter. To respect the fact that the expression of this difference in the lives of actual women is *always inseparable* from other identifications such as race, class, ethnicity, language, religion, and sexual orientation, I will speak of *the feminine within the imaginary domain* rather than gender.

The crippling suffering of the storyteller is real enough and illustrates how much of a struggle it can be for some women to claim themselves as actively desiring subjects. Nevertheless, skeptics of psychoanalysis justify their suspicions for many different reasons, one of which is the rejection of the idea that one can be held captive by drives generated by the unconscious. The rejection of the unconscious often reflects an empiricist bias. This bias passes as common sense nowadays and holds that what we desire, we desire. It is as simple as that. But Cardinal's novel tells a different story of how thwarted we can be by unconscious drives that often make themselves known through bodily symptoms. The narrator—a version of Cardinal herself that shifts as she changes her position in time with respect to her own story—remembers herself completely captured by her symptoms. Cardinal was married when she began to bleed without relief in her late twenties; she already had three children. Sometimes the bleeding was so severe it became a serious hemorrhage. When the bleeding first began, though, Cardinal believed that she was suffering only from abnormally long menstrual periods or some irregularity in her cycle. If this was the case, Cardinal felt she could still exercise some control over her body. The control took the form of stopping all physical activities. If she kept still, perhaps the blood would stop. But her attempt to reign herself in gave way to panic before the onset of the untimely bleeding. Her time was being overrun by an uncontrollable "Thing." There seemed to be no form of self-constraint that would effectively curb the blood. Hours, sometimes days, of relief were all Cardinal's efforts of restraint would grant her:

> How not to speak of my joy on the days when it seemed to dry up, only to show itself in its brownish traces, brownish, then ochre, then yellow. On those days when I wasn't ill, I was able to move about, to see, to get out of myself. The blood was finally going to creep back into its tender sac and stay there as before for twenty-three days. With this end in mind I used to try to exert myself as little as possible. I handled myself with the greatest possible

precaution: not to hold the children in my arms, not to carry the groceries, not to stand too long before the stove, not to do the laundry, not to wash the windows. To move in slow motion, quietly, so that the blood would disappear, so it might stop its horrors, I would stretch out with my knitting, while watching over my three babies. Furtively, with a gesture of the arm, quick and adroit from habit, I went to check my condition. I knew how to do it in any position, so no one would notice it. Depending on the circumstances, my hand would slip down to my pubic hair, tough and curly, to find the warm, soft, moist place of my genitals, only to quickly take it away again. . . . And what if there was nothing there? Sometimes, there would be so little on it that I had to scratch hard with my thumbnail the skin of my index finger and my middle finger in order to produce an almost colorless sample of the secretion. A sort of joy would come over me: "If I don't make even the smallest possible movement, it is going to stop completely." I would be motionless, as if asleep, hoping with all my strength to become normal again, to be like others. Again and again I made those calculations at which women are so adept: "If my period ends today, the next will be on. . . . Let's figure it out. Does this month have thirty days or thirty-one?" Lost in my calculations, in my joy, in my dreams, until that strong and precise caress, very secret, very tender, would surprise me with a clot carried along by the blood.[22]

Cardinal lived through what she experienced as torturous, shameful incidents. The bleeding would begin as a ferocious flow, catching her off guard. Furniture would be ruined in her own house and in the homes of her friends. In desperation, she went to doctors who performed curettages to stop the bleeding. The bleeding would not stop. Nothing helped. The blood took over her life. There was nothing left for her but to surrender to the power of the blood. She finally stopped socializing or even leaving her house at all:

> What woman would not have been driven insane to see her own sap run? How could one not be exhausted by the surveillance without respite of this secret spring, nagging, observable, shameful? How to avoid using the blood to explain that I could no longer live with others? I had stained so many easy chairs, straight-back chair, sofas, couches, carpets, beds! I had left behind me so many puddles, spots, spotlets, splashes and droplets, in so many living rooms, dining rooms, anterooms, halls, swimming pools, buses, and other places. I could no longer go out.[23]

Cardinal's family became increasingly hopeless about her response to the bleeding. As a result, they took her to "experts." The exams made her feel

"raped" and added to her sense of intense shame. Finally, one doctor diagnosed her with fibroid tumors and scheduled a hysterectomy. Only thirty at
the time, Cardinal insisted on keeping herself physically intact. Her effort
failed: the flow encompassed her life and she finally suffered a psychotic
breakdown.

To tell her own story, Cardinal divides herself into two people, the one
before and the other after her psychological birth. Yet these two are one, intertwined in the woman who writes the history of the psychosis she endured
and her struggle to free herself from the "Thing" that had captured her:

> I must think back to find again the forgotten woman, more than forgotten,
> disintegrated. She walked, she talked, she slept. To think that these eyes saw,
> that these ears listened, that this skin felt, filled me with emotion. It is with
> my eyes, my ears, my skin, my heart that that woman lived. I look at my
> hands, the same hands, the same fingernails, the same ring. She and I. I am
> she. The mad one and I, we have begun a completely new life, full of expec-
> tations, a life which can no longer be bad. I protect her; she lavishes freedom
> and invention on me.[24]

The recovered writer describes the "mad" woman in the last weeks and
days before she is finally hospitalized. To survive at all, she has to rely more
and more on tranquilizers. But they bring only intermittent relief from her
waking nightmare.

> It was between the bidet and the bathtub where she felt most secure when
> she could no longer master her internal functions. It was there where she
> hid while waiting for the pills to take effect. Curled up like a ball, heels
> against buttocks, arms holding the knees, strong, tight against the cheek,
> nails dug so deep into the palms of her hands they eventually pierced the
> skin, her head rocking back and forth or side to side, feeling so heavy, the
> blood and sweat was pouring out of her. The Thing, which on the inside was
> made of a monstrous crawling of images, sounds, and odors, projected in
> every way by a devastating pulse making all reasoning incoherent, all expla-
> nation absurd, all efforts to order tentative and useless, was revealed on the
> outside by violent shaking and nauseating sweat.[25]

Finally, Cardinal's family acts definitively and puts her in a sanitarium. The
sanitarium is not for the mentally ill because the family does not want anyone
to know of her true illness. As Cardinal describes it, madness is not acceptable
in the families who have only recently acquired middle-class status. Cardinal's

mother inherited a small, profitable vineyard in Algeria from her own mother. But their status in French colonial society was tenuous. Her mother was always worried about losing her money and her class status. Cardinal's family placed her under the care of an uncle who sedated her and kept her condition a secret. As the hospitalized Cardinal saw it, the idea was to use medication to turn her into a shell of herself. She could then be made presentable for polite middle-class society since she had been gutted by the medicine. Not formally hospitalized, Cardinal simply walks out of the hospital and goes to a friend's house. The next day, she visits a psychoanalyst with whom she stays in treatment for the next seven years. Cardinal would never remember whether it was her friend or herself who made that first appointment. The analyst told her he agreed she should stop her medication altogether for the treatment to proceed. The Cardinal at the beginning of analysis only vaguely grasped the possibility of being otherwise: "But at that time, I didn't know that I had hardly begun to be born again and that I was experiencing the first moments of a long period of gestation lasting seven years. Huge embryo of myself."[26]

A Psychoanalytic Conception of Dignity

So far, I have written of Cardinal's story of her "mad" self only to reinforce how profoundly someone can be imprisoned by her psychic life, so much so that it makes it impossible for her to reach into her own desires, let alone express them to others and act on them in her life. But I now want to shift gears and rely on the story of her struggle to free herself of her crippling symptoms. Her analytic work will illuminate how psychoanalysis serves the dignity of the desiring subject.

At the time of her first meeting with her analyst, Cardinal, the recovered writer, remembers that her earlier self had some vague idea of why she had collapsed into psychosis. Her parents were divorced. Her father saw her rarely. The lack of regular contact with him made him seem strange and his world impenetrable. Her mother always described her marriage as a mistake. The spoken reason for the divorce was her fear that her husband's failing business would take them all down. But early on, the traumatic death of the couple's first baby shook the foundations of the marriage. At that time, Cardinal's mother thought seriously about divorcing her husband. Because the baby had died of tuberculosis, the mother blamed the death of the baby daughter on him. Until the death of her daughter, Cardinal's mother had no idea of her husband's tuberculosis. She always condemned her husband for "killing" her baby; had she known of his illness, she would have tried to protect her daughter. Ostensibly to take care of her baby and get medical attention for her, the husband had sent Cardinal's mother away from Algiers. As a result, Cardinal's mother was on her own in a hotel room when the baby died. She was twenty years old. The trauma of the loss dominated her life. Cardinal remembers how her mother's grief would frequently take her over, how the whole house became still before her anguish:

> Often I would hear her whining from the bedroom. Through her door would come little noises, the rustling of tissue paper mingled with faint sobs, and sometimes a lament: "Ah, my God, my God." I knew she was on her bed, unwrapping the relics of my dead sister: slippers, locks of hair, and baby clothes. Nanny would then behave as though she were in church. She would cross herself, mumble prayers, her eyes suffused with tears.[27]

Cardinal's mother was unapproachable during these episodes of uncontrolled weeping. Cardinal knew that her mother had been burdened by her pregnancy with her. This pregnancy was discovered only after the divorce papers had been filed. The divorce extracted a pound of flesh from her mother who, being a devout Catholic, wanted the Church to recognize the legitimacy of the divorce, but knew that, in doing so, the Church would demand that she never marry again. Her husband's life, however, could go on. Her mother portrayed him to Cardinal as wayward, unreliable, and not truly a middle-class gentleman. His allure was associated with sin, illness, and shadowy undertakings. Cardinal was supposed to keep her distance, but she was also put in the position of being her parents' go-between. She was commissioned by her mother to bring her father's support check back home whenever she visited him. Anxiety about making sure she would get the check colored her visits to her father. He died from tuberculosis while Cardinal was still a young woman:

> For me, Father is an abstract word without meaning, since Father goes with Mother, and those two persons in my life are unconnected, far from one another, like two planets obstinately following their different paths in the unchanging orbits of their separate existences. I was on the mother planet, and, at regular but distant intervals, we would intersect the path of the father planet, shrouded in its unwholesome nimbus. Then I would be ordered to shuttle between the two, and, as soon as I had set foot again in the realm of the mother, as soon as she had retrieved me, she seemed to accelerate her course, as if to carry me away more quickly from the ill-fated father

planet . . . I know that I know nothing of the paternal side of men, if indeed there is one.[28]

As Cardinal's analysis proceeds, she begins to see how she imagined that the only bond between herself and her mother was death. That imaginary bond was her "internal tyrant." As a young girl, Cardinal saw her mother as obsessed with death and with ceremonial rituals associated with the dead and the dying. That girl frequently accompanied her mother to the resting site of the baby who would have been her sister. Her mother tended the grave, which had to be perfect for the baby who had died before it could ever fail her mother. Cardinal believed she had always failed. And her mother spoke to the dead baby. In Cardinal's imagination, her mother had taken the dead baby back inside of her again, and longed deeply for another chance with it. As Cardinal watched her mother caressing the tombstone, she wished herself in her dead sister's place in order to receive the devotion she felt she had never been given:

> At those moments I would have loved to be the stone, and, by extension, to be dead. Then maybe she would love me as much as she did this little girl I had never known, and whom, it would seem, I resembled so little. I saw myself stretched out among the flowers, ravishing, inert, dead, and her covering me with kisses.[29]

The recovered Cardinal recounts a number of breakthroughs throughout her treatment. She brilliantly illuminates the analytic process by recounting how the free association with the word "tube" takes her back to a scene with her father who sees her urinating, and to another in which she masturbates with the toilet-paper tube in the bathroom. The breakthrough does not simply arise by remembering what shame had repressed. By reconnecting to her delight and her early expressions of sexuality, Cardinal began to claim her own desire. Less than a year into her analysis, her bleeding stops. Freed, then, from the grip of the flow, and in touch for the first time with herself as an actively desiring being, she moves on to the next stage of her analysis: "During the first part of the analysis I had won health and the freedom of my body. Now, slowly, I was going to begin to discover myself."[30]

Cardinal the writer remembers in her narration how she finally came to understand the blood as her body's way of acting out her mother's explicit wish to abort her. On the level of the imaginary, she was dictated by an unconscious command: "Be aborted. Bleed yourself away." It was not until she had reached adolescence that Cardinal came to know the whole story when her mother finally told her about ending her pregnancy. Her mother had also told her the meaning of becoming a "young lady" in that same conversation. Her mother's own experience blends into her explanation of why Cardinal's free life with her male playmates—sons of the workers in the vineyard—had to change: once a girl has her period, it is no longer safe to be around boys.

Cardinal first remembers her conversations with her mother about babies and menstruation happening in a highly solemn fashion in the family living room. She associates her mother's discussion of feminine "hygiene" with her mother's ladylike behavior, reimagining the discussion as having happened in such an appropriate setting for ladies as during tea time. But in the course of her analysis, she is brutally hit by the memory of where the conversation actually took place. In fact, it had happened while walking along a downtown street in Algiers. It was on the street that she was told by her mother just how "stuck" a woman could be once she is invaded by the presence of a fetus within her womb. The recovered writer sympathetically describes how her mother could not possibly dream of an abortion because of her religion, and how she could not envision herself seeking help from "evil" women doctors who performed such operations. The adolescent Cardinal, on the other hand, could hear only how her mother did everything she could to "rid" herself of the fetus "naturally." She deliberately rode horses for hours, trotting and galloping so as to jolt the fetus out of her. She took long bike-rides and played tennis during the hottest time of the day. But no matter how vigorous the physical exercise, she was unable to induce the miscarriage. She even swallowed quinine and, on several occasions, took aspirin by the bottle. But nothing happened. Cardinal remembers her mother's recollected fury at the thing within her as she told her daughter about her futile attempts to put an end to her pregnancy:

> "Listen, well: when a baby has taken hold there's nothing you can do to dislodge it. And you get pregnant in a matter of seconds. Do you understand me? Do you know why I want you to get the benefit of my experience? You see how one can be trapped. Do you understand why I want you to be forewarned? Do you understand why I want you to know that you can not trust men? . . . After more than six months of treatment, I had to resign myself to the obvious."[31]

Even Cardinal's birth was retold numerous times throughout her life as particularly difficult, the labor pains being much worse than those with the other children. Cardinal was not properly positioned and she was born face up, which made her cheeks turn red due to the pressure. Retelling it, Cardinal's mother laughed at the memory of the first time she set eyes on this "made up" baby. In the course of her analysis, Cardinal realized she had heard, and had been responding to, an unconscious message from her mother. She echoes the message:

> Go away, you little shit, get the hell out of here! . . . Still moving? Here's something to calm you down. Quinine, aspirin! Sleep, little darling, sleep, little baby, let me rock you; drink, my beauty, drink the lovely poisoned brew. You'll see what fun you're going to have in the toboggan of my ass when you're well and truly rotted by drugs, drowned like a sewer rat. Death to you! Death to you![32]

Cardinal remembers that in the middle of her analysis, she finally confronts her mother's "beastliness," a condition she names "not because she wanted an abortion . . . on the contrary, her beastliness consisted in not having followed through on her desire to have an abortion."[33] At that time, she also sees how she became the remains of her mother's trauma as her own womb bled.

Given the divorce, her mother's life during the day was devoted to restoring her status as a proper upper-class lady of the colonies. She was rarely at home since her days were devoted to nursing the "natives" and their children. Without her "Algeria" she was nothing. The fantasy of herself as the one who cared for the natives sustained Cardinal's mother. The revolutionary uprising of the Algerian people for national independence drove her and her family out of the country. Cardinal says little about her and her mother's relationship to Algeria, but she does describe again and again how her mother's administering care to the "natives" was her way of surviving. Even the recovered writer rarely speaks about how she was psychically affected by her own standing as French in Algeria. In a revealing comment, however, she writes: "It seems to me that the Thing took root in me permanently when I understood that we were to assassinate Algeria. For Algeria was my real mother. I carried her inside me the way a child carries the blood of his parents in his veins."[34]

Even the recovered writer continues to remember her relationship to Algeria as her real mother. She points to a lingering unconscious identification and to their status as white colonial women in Algeria. The *persona* of the martyr was one offered by her mother's class position as a white upper-class colonial. It was a mask and a masquerade that required "the natives" in order for her to make sense of who she was. This was what sustained her mother. Cardinal's mother could not ultimately survive once she was stripped of her mask by the Algerian revolution. The Cardinal who ends her book as a feminist still has not grasped the place and meaning of colonialism, neither in her own life nor in her mother's. Once she left Algeria, her mother degenerated further and further into alcoholism, and lost the *persona* that had previously sustained her.

Perhaps unconsciously, Cardinal lets the reader know that her mother always had a drinking problem as she describes her slowly enjoying a glass of wine. I write "unconsciously" because even Cardinal the writer does not seem aware of how her mother's alcoholism affected their relationship, nor why it became life-threatening once she left Algeria. Ultimately, both Cardinal's mother and her grandmother moved to France. During the final stages of her alcoholism, after her own mother's death, Cardinal's mother moved in with her daughter. When Cardinal found her drunk and sitting in a bed befouled by urine and excrement, she decided she was unable to take care of her mother and told her mother that she could not live with her anymore. Shortly thereafter, her mother died of alcohol poisoning. Only after her death does Cardinal's mother rise before her as a person. At first, her mother's death came with a sense of freedom, even though she was constantly haunted by her mother's face: a horrible death grimace. But then, as Cardinal writes, "Something wasn't right, however. I didn't feel as free as I said I did."[35] With her mother in the grave, Cardinal finally finds the words to tell her that she loves her. She no longer blames her mother for the "savagery and butchery" that lived between them:

> If I had not become insane, I would never have emerged. As for my mother, she had forced back her insanity until the end, until the departure from Algeria. It was too late, the gangrene had gone into her marrow. She was afraid to rebel through the words and the gestures of rebellion, she did not know them, THEY had never taught them to her.[36]

Competing Analytic Stories: Cardinal's Collapse into Psychosis

There is a classic psychoanalytic interpretation for the illness endured by Cardinal that was once accepted by many analytic schools and perhaps still is by some. Cardinal could not escape what she called the "mother planet,"[37] with all its grisly relationships and overwhelming despair. There was no father present enough to supply her with other identifications that could provide her with an ego ideal strong enough to stand up to her mother's pull into a psychic world dominated by death. The bitter divorce made her visits to him irregular, and they were tainted by her mother's anxiety and disapproval. Having the responsibility of getting the alimony check from her father made her anxious. It reminded her that he was unreliable; he would seldom do what he should for her, his own daughter. His world remained mysterious and eerie. Although he was clearly affectionate and wanted a relationship to exist between the two of them, he could not exert enough force to pull her onto the "father planet" and thus allow her to differentiate herself from her mother. Once he died, he left only scary, ghostly memories, which she could not decipher. She did not know who her father was or what he represented. She was ultimately pulled into her mother's downward spiral of guilt and depression. As Bruno Bettelheim notes in his afterword,[38] her mother could only drag her through the haze of her guilt, not only over the lost baby who had died, according to her mother, because of her foolish marriage to an irresponsible husband, but also because of her attempt to abort Cardinal. Bettelheim speculates that her mother's attempt to fix Cardinal was a defensive compensation for her fear of having somehow damaged Cardinal in her efforts to rid herself of an unwanted child: a perfect little girl would assuage her guilt. But there is no such thing as a perfect little girl. Cardinal remembers both her desperate efforts to fill her mother's expectations and her constant failure to do so.

But then, she seemed destined to fail because she unconsciously identified herself as a sign of the failure of her mother's own body, of her "sex" that had functioned as her enemy and against her will. Without any countervailing forces in her life to extricate her from her identification with her mother's conscious and unconscious wishes, Cardinal succumbs to them. She ultimately acts out her mother's disgust with her "sex," doing this just as her mother is taken over by her own body. Her womb betrays her too. By bleeding uncontrollably and coming closer to death, she meets her mother's own wish to die. At twenty-seven, her mother became pregnant with the unwanted Cardinal. At the same age, the daughter starts suffering from continual menstrual bleeding. Bettelheim suggests that Cardinal was able to hold on to her sanity throughout her childhood because of her mother's distance from her. Due to the demands of her charity work, she was gone almost every day. At night, alcohol was her companion. Cardinal once watched her mother overtaken by dreaminess as the effect of the wine soothed her pain and lifted her despair:

> At the end of the corridor, I saw her bathed in light, which was the more dazzling in contrast to the deeper shadow in which I found myself. In her hand a large glass of wine. She was motionless, sad, and calm. She looked far away, very far away. She would drink in great gulps, sometimes, closing her eyes. I had the impression that it did her good. The glass emptied, she went into the semidarkness of the pantry, opened the refrigerator, which gave off a gay and reassuring light, took out a bottle, filled her glass, turned out the kitchen light, and then, feeling her way, went towards her room, her viaticum in her hand. She locked her door. I knew she wouldn't stir from there until the following day.
>
> While she was alone in the light of the kitchen, I saw her drink her white wine and I wanted to be the wine, to do her some good, to make her happy, to attract her attention.[39]

For Bettelheim, it was an ironically saving grace for Cardinal that, although she desperately tried to hold on to her, her mother was much too adrift within her own despair to be able to cling to her in return. There was, however, room for a third person, a loving nanny, to enter as a go-between to counteract the coldness of the mother toward the daughter. Bettelheim also suggests that the mother's honesty about her attempt to abort her daughter gave Cardinal a powerful reference point for her intense feeling about her mother's withdrawal from her. Her mother's frank, even brutal, account allowed Cardinal to *know* that it was not all in her head. Because Cardinal cannot reach out for her mother when her alcoholism finally becomes life-threatening, the mother's power over her daughter is never truly broken. Bettelheim sees this as a failure in her analysis. But he explains that because of the power of parents over children, and because of the seriousness of Cardinal's own illness, it is not surprising that she could neither free herself completely from the fear of her mother's stranglehold over her, nor free herself enough to reconcile with her mother in life and feel capable of caring for her during her last days.

In his preface, Bettelheim writes, "Reading this book leaves one with great admiration for the author and restores one's faith in man."[40] As for me, I was left with admiration for the author, but also with overwhelming grief for the mother named and described at last at the end of the book: Solange de Talibac, the redhead with the green eyes. As she died, she left behind a death-mask of extreme anguish.

Solange de Talibac did not seem to have any inner compass to direct her life. Instead, she seemed directed by her station in life, by her position as a Catholic, as a "white lady" and a landowner in a French colony. She had no alternative directions for her daughter concerning how to become a woman. Besides those conventions, she passed on her hatred and fear of her "sex." Cardinal vividly describes the meaning of having a vagina as a metaphor for this specific feminine terror:

> I had begun to think, as never before, of what it meant to be a woman. I thought of our bodies, mine, my mother's, the others'. All the same, all having holes in them. I belonged to that gigantic horde of penetrable beings, delivered to the invaders. Nothing protects my hole, no eyelid, or mouth, or nostril, or grating, or labyrinth or sphincter. It hides in the hollow of soft flesh which does not obey my will, and is naturally incapable of defending it. Not even a word to protect it. In our vocabulary, the words which designate this particular part of the female body are ugly, vulgar, dirty, coarse, grotesque or technical. . . . Could this be the origin of the essential fear as old as humanity, unconsciously submitted to and forgotten? A fear which women alone could feel and transmit instinctively, which would be their secret? A fear which would be attributed to violent penetration by men, but which would in fact be much more deep running and profound. A fear invented by women and taught to them by other women. Fear of vulnerability, of an absolute inability to shut ourselves off completely.[41]

The fear of "having" a vagina stands in for the lot of women, "the red of her blood, the black of her fatigue, the shit brown and pus yellow of diapers and underpants worn by her babies and her man. And then the gray weariness and the beige of resignation."[42]

Cardinal was delivered from the venom too late to salvage a relationship with her mother. Their shared "sex" remained a barrier between them because "it" was the thing that imprisons the soul. The tragic irony in the title of her book—*The Words to Say It*—is that Cardinal never found the words that would allow her to reconcile with her mother while she was still alive. It is perhaps the deepest and saddest truth that Cardinal could not call her mother by her own name. Neither mother nor daughter could find their way to voice different possibilities for their "sex."

Cardinal writes that her mother never taught the words that would enable her to rebel against the seeming destiny of her "sex." But does Cardinal herself have the words at the end of her account of her own analysis? We know from her life that after the publication of her book, she became a prominent feminist activist. Feminism was her fate. She had to keep struggling to find the "words to say it"; to speak the absence which led her mother to "act out" her unspeakable despair with her "circus revolver."[43] By the end of her analysis, Cardinal clearly connects her mother's tragic life with the patriarchal culture and society in which she had to survive. Only by challenging these as a feminist could she claim for herself the freedom her mother could not imagine.

A psychoanalytically informed feminism creates the space for a new future only by working through the past; only by allowing us to remain haunted. By confronting the "THEY" who robbed Cardinal's mother of "the gestures of rebellion," we are able to envision a return to the future. With the introduction of the "THEY," we no longer speak in purely private terms, nor in idiosyncratic family dramas. But who is this "THEY" who orders sexual difference?

Why Lacan?

In Jacques Lacan can be found the concept of a Symbolic Order into which human beings must enter if they are to achieve psychic life. This idea of a Symbolic Order can serve as a starting point in the effort to come to terms with both the impersonal character and the peculiarly public nature of the "THEY" who orders sexual difference. Lacan's Symbolic Order is public in a peculiar way, because, even though we are all subjected to it, this Symbolic Order is not a space where we enter together to consciously shape our institutions and laws. The Symbolic *orders* us, not the other way around. For Lacan, the Symbolic Order is not identified with any set of culturally specific prohibitions and social institutions. Nor is it only another way of noting that human beings are as they are simply because they speak and are thus formed by the very language that allows them to articulate who they are, how they live in history and struggle within social institutions. At least during one period of Lacan's work, the Symbolic is *the law of inculturation* that inscribes the embodied human being as a sexually differentiated subject.[44] This sexually differentiated subject further owes her subjectivity to such psychic laws as the incest prohibition. She is not the master of these laws.[45] These laws enforce symbolic castration, which ensures that the separation of the child cannot be encompassed by the mother's world.

This law of symbolic castration would seem to be true for both sexes. Understood at the level of a fundamental law of culture, this law does subject all of us to it. Yet, at least at one stage of his thinking, Lacan seems to maintain that the law of symbolic castration, which makes us the *human* creatures we are, does not simply sexually differentiate us. It enforces a particular kind of sexual differentiation. We are thrust into a dichotomy: we are either masculine or feminine, man or woman. In Lacanian terminology, there are two positions that human culture offers us: either we are identified and thus come to identify ourselves as the one who *has* the phallus or the one who *is* the phallus. There is no direct correlation with biological sex. No matter what our actual body parts might seem to dictate, any of us can take either position.

These two positions, however, are not symmetrical. There is a symbolic referent for the masculine position: it is the one that *has* the phallus. The masculine position allows little boys to place themselves in the line of paternal descent. At least theoretically, identifying as a man gives the little boy a compass to direct himself as an active subject. He finds himself in the metaphors that identify subjectivity with masculinity. What Lacan means by this is not really so hard to follow. Even if we disagree with his conclusions about language as the guarantor of sexual dichotomy, we can all think of day-to-day examples that make it clear what he means. A man "stands up for himself," "holds the line"; he does not "beat around the bush" or "skirt the issue." No "softy" is he. To urge a male to do the right thing is as comprehensible as to command him, "Be a man." Many writers have written that men in the United States have been "stiffed" because those metaphors are not as often present for them as they previously were.[46] But the metaphor of being "stiffed" implies that men feel they have lost something that is rightfully theirs: the father's command to "Be a man" should really mean something and thus guide them into adulthood. The loss of such metaphors, qualified as "stiffed," inadvertently shows how important they have been and continue to be for contemporary masculine identity. This is why many men not only feel "gypped" but also feel entitled to complain about it.

If a woman wished to guide her daughter by simply saying "Stand up and be a woman," would that be equally comprehensible? Is it not a mixed metaphor? Does a woman "stand up" as a woman only by identifying herself with the attributes of the masculine? What other meanings could this command carry: "Be compassionate"? "Dress better so you can be sexually attractive"? "Learn to cook, because the way to a man's heart is through his stomach"? "Become someone with the attributes to be a loving mother"? We now know what the normative meaning of *becoming a woman* was for Cardinal's mother: "All the fun is over. From now on you had better watch out. But eventually THEY will get you anyway." For her, it would have been a contradiction in terms to command "Stand up and be a woman." Being a woman meant "lying down" for THEM. Because of the Catholic Church's demand that she never marry again, and because of the experience of marriage and pregnancy, she repudiated the most literal meaning of "lying down" for THEM, but never "stood up" for herself.

Cardinal's mother assumed another acceptable *persona* of white upper-class femininity, and she ultimately collapsed into alcoholism. Lacan's point is that it is not a mere coincidence that there is no obvious meaning to femininity. At one point in his life, Lacan would have had to agree with Simone de Beauvoir that Man stands in for the human whenever the human is identified as acting subject.[47] When Lacan writes, "Woman does not exist,"[48] he means that the meaning of femininity is left to float precisely because "it" cannot fit into the Symbolic Order.

Femininity is the masquerade that women play to represent themselves as the object that can fill man's desire.[49] When Lacan writes that Woman is the symptom of Man, it is because she is only presentable to him as the object of his desire. Man desires. Therefore, there must be someone for him to desire. There is Woman who appears to be what he needs to fulfill himself. This Woman is nothing but a fantasy of what he does not have, and yet conjures up, as the object of his satisfaction. If we simply release her from her entrapment by the man's fantasy, there is no essential Woman underneath this fantasy who can be reached. There are a few other feminine personae such as the one assumed by Cardinal's mother, but none of them are consistent with standing up for oneself as a *woman*.

According to one reading of him, Lacan is rightfully infamous in feminist circles, not only for his elaboration of a law of symbolic castration, which holds for both sexes, but for his identification of symbolic castration with the paternal law, often mistakenly conflated with the law of the phallus.[50] This erroneous reading by some feminists still leaves us with the question of whether Lacan's idea of symbolic castration, as a necessary psychic law through which we enter culture and become human, is really only an ideological legitimation of patriarchy. To be sure, according to some of the received interpretations of Lacan, Cardinal's collapse into psychosis would be attributable to the absence of her father and to his early death. When he was present in her life, he was not allowed to assume the symbolic role of the father, and thus he was unable to enforce the paternal law that would have freed Cardinal from the "mother planet."

Yet, even under this interpretation, Lacanians are referring the desire of the mother to the symbolic father. The actual father *matters* only when the mother's desire for him is either strong enough or, on the contrary, too weak for the man in the woman's life to represent the paternal function as a metaphor of the need for psychic separation. In modern Western patriarchal societies, the reference point for the heterosexual mother's desire often remains the woman's man or men. Thus, some Lacanians lapse into conflating the paternal metaphor with the actual father because the role of the father, if he is the referent of the mother's desire, matters in the child's ability to achieve separation. Clinicians in particular often analyze the child's dilemma in a heterosexual patriarchal context. Their concern is usually with their patient, and not with the broader social and political context in which the patient lives. This may explain why, in analytic accounts, the conflation of the actual father with the function of the paternal metaphor occurs. Even so, at the level of theory, Lacanians always distinguish the actual father from the symbolic function of the paternal metaphor. However, when Lacanians fail to remain "true" to the significance of this distinction, when they literalize Lacan by conflating fathers with the paternal function, they do not differ much from American ego psychologists. They can also be less subtle—in the analysis they would offer of Cardinal's illness—than Bettelheim, who at least sympathizes with her mother's plight and efforts to warn her daughter through her brutal descriptions of the dangers of femininity.

If this conflation is made, then, even if Lacan can help us understand the peculiarly public nature by which we are sexed—which could help us explain why the psychic process we undertake in order to be sexed is not something to be understood as a simple, private affair—feminists could still rightfully reject his offering us one more justification for the necessity of patriarchy. If the human subject comes into being only once he or she passes through certain psychic laws, inseparable from the presence of actual fathers to impose them in the family, then public institutions should arguably be rendered consistent with those processes. This kind of thinking was clearly crucial to the opposition posed by some Lacanians to the suggested legal reforms in France that would grant gay and lesbian parents' parity with heterosexuals. Their argument was simple: without heterosexuality there will be no father present to assume the paternal function, thereby leading children of gay and lesbian parents to collapse into psychosis. By this account, the role of psychoanalysis is to enforce the role of the father in clinical practice, and, when necessary, to do so within an actual legal system.

Gurewich's Interpretation of Lacan

I will now turn to the work of Lacanian clinician Judith Feher Gurewich, who is always careful to distinguish the paternal function from existing fathers in any given family setting. She can help us see the significance of Lacanian psychoanalysis for feminist theory. She interprets the meaning of symbolic castration and its relationship to the incest prohibition and the Oedipal complex. For Gurewich, it is both a clinical and an ethical mistake to invest psychoanalysis in the enforcement of the Oedipal myth as classical psychoanalysis perceives it. Instead, the goal of psychoanalytic treatment is to dissipate the destructive effects of the Oedipal fantasy and, with that effort, play a role in dispelling the hold of patriarchal definitions of masculinity and femininity. I turn to her precisely because of the connection between feminism and Lacanian psychoanalysis that we can glean from her work. Once we clearly understand that Lacan's notion of symbolic castration and Sigmund Freud's understanding of the incest taboo function according to different registers, we can, with Gurewich, reinterpret the ethical goal of psychoanalysis. The result of this demonstration is that the Oedipus myth becomes an effect rather than a cause of human subjectivity. Gurewich raises and seeks to answer the very question that feminists have demanded be answered: How are we to use Lacan to help us explain and understand why our existence as actively desiring subjects cannot be taken for granted and relegated entirely to the private realm? To quote Gurewich: "If Lacan scrubbed Freud's discovery clean of its biological underpinnings isn't it time to apply Lacan's scouring pad to his own reading of Freud, specifically where the shadow of patriarchy obscures the distinction between the desire of the subject and the demands of cultural ideals? In other words, can Lacanian psychoanalysis resist Lacan's own ideological or cultural bias?"[51]

By reexamining Lacan's later conception of the *jouissance of the Other*, she effectively uses the "scouring pad." This can help us rethink his earlier formulation of the mirror stage and the Imaginary in its relationship to the Oedipal myth. But first, we need to look at exactly what is meant by the *jouissance of the Other*.

To distinguish it from desire, Gurewich gives us a succinct definition of how Lacan uses the word *jouissance*:

Let's note right away that the term *jouissance* cannot be easily translated by enjoyment. *Jouissance* is a legal term referring to the right to enjoy the usage of the thing, as opposed to owning it. The *jouissance* of the Other, therefore, refers to the subject's experience of being for the Other an object of enjoyment, of use or abuse, in contrast to being the object of the Other's desire. The distinction between desire and *jouissance* should at this point become apparent. The experience of being perceived as an object of desire implies that the individual can figure what it is about him or her that can be attractive to the Other. The desire of the Other in that sense offers the subject clues to what it would take to behave or to be what the Other wants. In contrast, the experience of being the object of the Other's *jouissance* conveys a sense of frightening mystery: What is going to become of me? What does the Other want from me?, etc. This is a situation in which the subject is clueless. The Other appears as enigmatic, as able to threaten the very core of the subject's being. The *jouissance* of the Other thus has the special quality of being both real and mythical. It is real, of course, in the sense that it informs the subject's psychic reality; it is mythical in so far as the power—the *jouissance*—that the subject attributes to the Other does not really exist. It is a necessary contingency of the process of psychic development. The goal of psychoanalysis therefore consists in demystifying the Other in its all-powerful and threatening incarnation.[52]

Her argument proceeds as follows: the child is hooked into the *jouissance* of the primary Other during the mirror stage. The mirror stage in Lacan is a structural moment in which the child comes to see herself as the beloved object of her mother. This structural moment is the Lacanian Imaginary. The child comes to see herself through the Other's eyes. The image of herself will be formed from the way she is seen by her primary Others, and this image in turn will shape the child's ego ideal. The ego ideal then directs the child's aspirations toward what she might become. In child rearing, this takes place on the simplest and everyday level. As her primary Other cheers for her early attempts to stand on her own two feet, the child sees herself as a "walker." But the experience of her mother's love, or of someone else who is in the place of the primary caretaker, carries with it a different message. The mother's love shows mother as a desiring being. After all, she loves the child and therefore is not completely self-sufficient. The child is someone who brings her something, someone who fulfills her. This experience of the mother's love also shows the mother as lacking, as wanting someone outside herself. As long as the child sees herself as the heart of her mother's life, she is spared the experience of her mother's *jouissance*. She remains at the center of her attention, and therefore her mother's desire is not threatening to her. But mothers are not just mothers. They are desiring beings. They leave. They go to work. They become obsessed with the play they are writing. They slip away for an afternoon of love. They stay late at a political meeting. Once the child runs up against *any* signifiers of the mother's desire, she begins to formulate a retroactive fantasy of what once was and tries to go back to that imaginary place where there is one Other just for her. The mother's desire is the barrier that keeps the child from going back.

For a Lacanian, this desire means that the mother is lacking something and, worse yet for the child, that her lack is not being effectively replenished by her. What she cannot fill becomes a gap that threatens the child. The mother's slipping away puts the child on the trail of the mother's desire: What is it she wants that is not me? As Gurewich explains:

> It is at this crossroads between the *jouissance* of the Other and the desire of the Other that Lacan situates the introduction of the prohibition of incest. The prohibition of incest is perceived by Lacan as the child's ability to identify with the clues, the "signifiers," the signposts of the mother's desire that can lead the child to a safe harbor usually provided by the desire and interests of the father.
>
> The signifiers of the mother's desire therefore provide limit to the mother's *jouissance* in the sense that they will propel the child towards new poles of identification through which the ego ideal will be constituted.[53]

In other words, the incest taboo signifies that the child is a separate being from the mother and cannot return to her fantasy of immersion with her. That she is a separate being from her mother and can survive as such allows her to go on and differentiate her own desire from her mother's. Notice that Gurewich says that the signifiers will usually lead to the father and then to a identification with his interests. In patriarchal society, it is in fact the father who is found as the outcome of the child's detective work in tracking the mother's desire. But Gurewich's point is structural. What the child must find to separate herself from her mother—and yet find her mother again as an actually separate human being who can continue to support her—is a way to symbolize her mother's desire. If the mother's desire is still something mysterious, if in her tracking quest there is no end that makes sense to her, she will remain trapped in the fantasy that she can still be everything to her mother. Her mother's desire must have a symbolic referent point beyond the child that she can understand. But there is no structural reason, at least under Gurewich's interpretation of Lacan, for that reference not to be either her lover Betty, or a play she is obsessed with finishing, or the printing company that she runs. On the other hand, if the mystery does not come to an end, the child remains trapped in the fantasy that she can still be the only one for the mother. Lacan identified this fantasy as the child's desire to be the mother's phallus.

The use of the word *phallus* has caused part of the confusion in some feminist criticism of Lacan. It is undoubtedly the case that Lacan often claimed that when the child discovers the lack in the mother—a lack that is the mother's *own* desire—the child will interpret such lack as the lack of the phallus. At a certain point in his work, Lacan argued that this identification of the lack in the mother as the lack of the phallus promoted the fantasy object of the phallic mother as one of many responses to the child's anguish over separation. The phallic mother has it all and therefore does not desire anyone other than the child.

In my interpretation of Lacan, the identification of the phallus as the imaginary object that the mother is lacking, and that she also seeks for fulfillment, is not a coincidence. This identification indicates the manner in which Lacanian terminology is drenched in patriarchal assumptions. The fidelity to these assumptions is the result of Lacan's determination to remain true to Freud's general conception of sexuality. Freud's concepts were developed through his interpretation of children, and his mistake was to see this imaginary as reality. He heard children or adults who recounted their childhood by reporting that they "saw" the mother as lacking the phallus. The mistake is to enshrine the child's imaginary within a patriarchal society as an unquestionable theoretical proposition. With his introduction of the concept of the Imaginary, Lacan, unlike Freud, was able to recognize explicitly that the mother's lack of the phallus is in the child's imaginary of her mother. The phallus is an imaginary object, which, once symbolic castration is fully assumed, is understood as such.

Neither sex has it. And the longing truly to have it, or really be it, is an expression of resistance to the symbolic castration, which proceeds through the signifiers that mark the separation from the fantasy that one can be everything for the mother and therefore never need to grow up. These signifiers are then repressed so that the child can divert herself from the track of trying to discover how she can satisfy her mother and get on with the business of her own life. Because each one of us undergoes this separation in our own way, these signifiers do not have any given content. When Lacan writes that the unconscious is the discourse of the Other, he means that we form a fantasy out of this original tracking of the desire of our primary Others in order to fend off our fear of being encompassed by the *jouissance* of those Others who leave us helpless before the mysterious demands they place on us. This fear becomes a wish, including the wish for the law that will keep us from going back and losing our subjectivity altogether. Thus, women too must be lacking because we can never truly be at one with the Other, nor can we attain complete satisfaction of our desire by turning someone else's subjectivity into something we can simply use. Because no one's fantasy is ever perfectly correlated with that of another, no one can ever be completely reduced to someone's ultimate object of desire. Our uniqueness resides in our fantasy life, and from that point of view, subjectivity is never completely extinguished, even in "madness."

For Lacan, the subject of desire is the subject of the unconscious. We lose the fantasy of being completely plugged into the Other, but with that loss we gain the world and ourselves as beings with our own psychic life. On the other side of psychic life is psychic death—the wish to extinguish the subjectivity that is born with our desire. Thus, Gurewich writes:

> *The wish for "incest" stands for the subject's desire to annul the division that has permitted the birth of his or her desire.* The pleasure principle is accessible only through the subject's identification with the signifiers of the desire of the Other. Lacan therefore locates the death drive on the side of this fantasy, in the sense that the fantasy speaks in some way of the death of the subject, the death of the desire that constitutes the subject.
>
> In turn, since this fantasy is experienced as a transgression of the law of the prohibition of incest, the need to keep it alive exposes the subject to yet another version of the death drive, the threat that comes form the superego. Castration anxiety and penis envy are therefore neurotic constructions that attempt to keep at arm's length a demand for a pound of flesh that the subject refuses to deliver. Hence the *jouissance* of the Other is not a force that preexists the subject. The wish for incest does precede its prohibition. The *jouissance* of the Other ultimately does not exist. It is a fantasmatic construction that carries at once a promise and a threat.[54]

The ethical goal of analysis is to reach the point at which the patient sees that the *jouissance* of the Other is within us and not an all-powerful someone who can actually obliterate us. By tracing the remains of the legacy of our separation from the maternal realm or from the world of the primary

Other, we can free ourselves from the fantasy that we cannot claim our desire because it is in the hands of an all powerful Other who forbids its expression. This fantasy is extremely difficult to release precisely because of the unconscious need to fend off the frightening *jouissance* of the Other by translating it into an unrealizable wish. What is encountered in the analytic process is the symbolic castration that allows us to separate our desire from the *jouissance* of the Other. Analysis exposes that this Other does not truly exist as a terrifying and incomprehensible object of *jouissance*. In Lacanian terms this other, too, is understood as lacking. We conjure up this frightening mother/Other from the signifiers we have attached to her desire, As it connects the mirror stage, or the Imaginary, to the Oedipal complex, this understanding of the *jouissance* of the Other allows us to free Lacanian psychoanalytic theory from its seeming need to promote patriarchal cultural ideals and the foundational heterosexual family, both of which attempt to ensure there is an actual father who assumes the symbolic paternal function. Instead, the work of analysis is to free the patient from the fantasy that she is living in the hands of the Other so that she can actively organize her life as a desiring being. What is crucial for Gurewich is that when the confrontation with the mother's desire occurs, the child must be able to find signifiers to reference what she wants to avoid: the trauma of remaining caught up in the world of the *imaginary mother* of early infancy. This fantasy mother is the maternal Thing that Cardinal believed entrapped her. Once the fantasy that the Thing could kill her had dissolved, what remained was the actual woman. But by then, Solange de Talibac had already died. Reconciliation with her actual mother before her death was no longer possible. Cardinal's analysis did not free her in time. This is the tragedy of the story.

Even while keeping in mind that Cardinal's mother was an alcoholic, Gurewich's reading of Lacan gives us another interpretation of the tragic story of Cardinal's relationship to her mother. As Gurewich writes:

> Children of alcoholic mothers are in my view, particularly prey to the fantasy of the *jouissance* of the Other. This is why I thought they would be good illustrations of what I am talking about, since they seem to have an unmediated relation to the lethal enjoyment that has taken over their mothers. In their case, castration, as Lacan calls the submission to the prohibition of incest, is made more difficult because they are excluded from the maternal realm not through the desire the mother has for the father, but through a sort of other worldly attraction for an artificial paradise where the father has

no place. The tendency to try to make sense of the mother's enigmatic *jouis-sance* therefore becomes overwhelming and compromises the child's ability either to identify with the father or to value sufficiently the love of a man who is impotent to satisfy his wife or to impose a limit on her addiction. Because the desire of the mother remains obscure, both self-love and desire will be impaired. The sight of the mother's attachment to an otherness that cannot be grasped hinders the children's ability to trust the true import of the prohibition of incest.[55]

Gurewich could be read to suggest that the problem is that the child cannot trace the signifiers of the mother's desire to the father, but her true point is a subtler one. Alcoholism takes the mother away from the child. At the end of the trail of the child's search for what the mother desires is something so other worldly that the child is unable to make sense of it. Cardinal's mother was not taken away from her by desire but by despair, a despair that could not be interrupted by Cardinal's attempts to please her mother. Her nights of drinking often ended in uncontrollable crying for her lost baby. She would lock herself away, but her sobbing and moaning could not be hidden. At those moments, it seemed to Cardinal that her mother's anguish filled the house. Even her daily treks out to the countryside to help "the natives" did not seem to be an expression of her own desire. It was the only way she could make up for her terrible "sins": the divorce and wanting her baby dead. The Catholic Church's command was that if she wanted her divorce to be recognized by the Church, she must not marry again.

Only after her death did Cardinal understand completely how her mother had been cut off from her desire. But to the child that she was, her mother's despair did indeed seem to be an abyss into which Cardinal could fall. She desperately tried to satisfy her mother, always failing and never knowing why. Cardinal presents a classic example of the severe impairment of image and desire that results from a mother who seems lost in some unknowable realm. All that seemed left to Cardinal was to join her mother in her drive toward self-obliteration. To be sure, she did not have a father who could offer her other poles of identification. But there might have been another road to psychic separation Cardinal could have traversed. As she ends her account of her analysis, Cardinal realizes that her mother did not have the words to speak her desire. Only then did Cardinal recognize that she had been completely caught up in what she projected onto her mother as her mother's *jouissance*.

I do not want to undermine Gurewich's clinical insight that in many cases patients in our own Western society—still patriarchal, even in spite of the extraordinary changes of the last thirty years—can be helped by identifying with their father's unconscious values. But what is structurally necessary is not the presence of the actual father, even though admittedly that may do the work of separating the child enough from the fantasy of being returned to her place as *the one* for the mother or, in Lacanian terminology, that she can be the mother's phallus. What is necessary is the symbolization of the mother's desire. After all, it is the *symbolic* mother—the one who desires—who at first challenges the truth of the child's fantasy of being in a dyadic relation to the mother.[56] This mother comes into shape for the child as long as the child can reference her mother's desire. In his later seminars, Lacan himself spoke of *names of the father* in the plural. These names were not for the actual father, but were for another Other—someone or an aspiration for the mother—that becomes the basis for separation.[57] There is no reason to presume that this second Other cannot be another woman. There is also no reason that it need be another person at all. It can be the articulated love of work, of writing, of politics, in which the mother names her own desire and the law that allows her to claim it.

This law is what I refer to as dignity. This law is what grants the mother her status as an actively desiring subject. It is not a law of her making, but it does indeed come with the territory of becoming a subject through psychic separation from the primary Other of infancy. That we must be subjected to the law that separates us so as to have a psychic life has come to be known as the *split subject*. We cross over into language and the social bond inevitably represented in language, and this passage becomes a one-way street because we can never give adequate expression to the fantasy thing that we have lost. My argument, however, is that there is no necessary reason to name this law of psychic separation "symbolic castration." This law is passed down to the child as the child's dignity as well. This law of dignity stands against her mother's encompassment of the child and also protects her mother's own desire. The law of psychic separation I am naming dignity recognizes that both mother and daughter are *in principle* beings who can become actively desiring subjects. A mother who stands up for her own desire, who claims it as her own, can split herself away from that fantasy Thing. By articulating her desire, by referencing it, she can pass down the one law that must be respected between mothers and daughters. It is this law of dignity that forbids incest, both literally and metaphorically. A child used by its parents, either in actual incest or in more subtle forms that encompass him or her as a "thing" in the parents' lives, will be ensnared in the *jouissance* of the *big Other*.

Kant's Insight

At first, this may seem a strange use of the word dignity since the concept of dignity is associated with Immanuel Kant's moral philosophy. For Kant, our desires are given to us by nature. As desiring beings, we are thus governed by the laws of nature. Our dignity, on the contrary, lies in our autonomy. As creatures capable of reason, we not only can value our own ends, but we can also discern which ends we should pursue on the basis of the moral law. There are several formulations of the Categorical Imperative, Kant's name for the moral law, but the essential idea is that we can be self-legislating if we follow a law that embodies our own free will. To exercise our free will is just another way of saying that we abide by the dictates of reason and by those alone, since everywhere else in our lives we are subject to the laws of nature, as are all other natural beings. To know whether we are acting from the moral law, we ask whether we can will our moral decisions as ones all other rational beings would accept as morally right, if they, too, were acting in accordance with their reason. We are free to the degree that we not only test our moral decisions in accordance with this law but also represent ourselves as acting solely on its basis.

There have been many critiques of Kant—and as many answers to those same critiques—that I do not want to address here.[58] Feminists in particular have argued that Kantian moral philosophy is both too individualistic and too rationalistic.[59] For some, this has led to the rejection of the ideas of autonomy, freedom, and dignity. My argument here is that psychoanalysis can help us reshape the ideas of autonomy and freedom, thereby salvaging dignity from a pre-Freudian understanding of desire. Our destiny as desiring beings is inherently social since we are produced as the unique subjects we are through our relations with the primary Others in our lives, who in turn are also shaped by the symbolic order into which they are thrown.

The *big Other* is what we fantasize as lying behind actual social conventions and which we unconsciously invest with the authority to govern us. Think again of Cardinal's mother, who invested the Catholic Church and the patriarchal conventions of French colonial life with absolute power over her, to the point that she was completely cut off from her desire.[60] All of us are traversed by our unconscious entanglements with our primary Others. But the ethical goal of psychoanalysis—to help us see that there is no absolute Other whose *jouissance* threatens us—can return our desire to us. Desire is born with our birth as subjects. Therefore, *in principle*, it is not something that can be taken away from us. In reality, of course, we might find it impossible to claim. But the fact that, *in principle*, we can claim it is the basis for our dignity and our freedom to undertake this struggle without being further hindered by outside forces such as patriarchal institutions. Because we can never reach the stage at which we draw a clean line between our conscious lives and our unconscious investments, we can never be completely independent of the unconscious. Cornelius Castoriadis offers a Lacanian-inspired definition of autonomy: "An autonomous subject is one that knows itself to be justified in concluding: this is indeed, true, and this is indeed my desire. Autonomy is therefore not a clarification without remainder nor is it the total diminution of the discourse of the Other unrecognized as such. It is the establishment of another relationship between the discourse of the Other and the subject's discourse."[61]

One reason for not being able to draw this line between the conscious and the unconscious is the structural moment of the imaginary, which is precisely a structural moment that can never really be surpassed. In the later Lacan, the Imaginary is identified with the ego; because it proceeds through identifications and our investments in the fantasy of the power of Others who initially shape our ideals, it can actually block the road to the unconscious and to our desire. Because it is a structural moment, the Imaginary is never something we can get rid of once and for all. This is why I believe there can be an alliance between feminist Lacanians and feminists in object-relations theory and intersubjective theory. By insisting on the need to protect the ideal aspect of ego ideals, both groups of psychoanalytic feminists give us a creative reshaping of the affirmative aspect of the Imaginary. But we can build this alliance only if we expand the Lacanian notion of the Imaginary so that it is irreducible to a structural stage neatly designated and clearly distinguished from the subject of the unconscious.

If we expand the notion of the Imaginary by giving it a potentially affirmative dimension—that is, by defining it as that which continues to feed the productive imagination—then we are also rendering the Lacanian concept of the unconscious more fluid. By using the Imaginary in all its richness to give us hypothetical fantasies about the meaning of these figures, we can even refigure the "internal tyrants" of the Imaginary. Cardinal's imaginary mother had to be exposed as such before Cardinal could ever reconcile with her symbolic mother. Unfortunately, though, her imaginary mother could only be dissolved in death.

Woman, with a capital *W*, is identified as the ultimate figure of desire, the phallic mother, who appears only in the repressed unconscious signifier of the actual mother's desire. The dissolution of Woman turns on Lacan's philosophical attempt to draw a neat line between the Symbolic and the Imaginary through an appeal to the Real of fulfilled desire, which is inevitably lost to us since it is never "there," except in our fantasies and in our words.[62] I agree with Castoriadis that the founding fantasies that individuate us always retain an imaginary dimension because there is no "self" there without this basic organization. In this sense, the Imaginary is a fundamental schema within which we struggle to know ourselves. Self-knowledge, in other words, must proceed through an endless reworking of this basic and yet ultimately imaginary schemata. To quote Castoriadis: "On the level of the individual, the production of this fundamental phantasy stems from what we have termed the radical imaginary (or the radical imagination); this phantasy itself exists in the mode of the actual imaginary (of the imagined) and is a first signification and scope of subsequent significations."[63]

We can know ourselves only as always already imagined. In that sense, there is a background we can never fully recapture. That residue of unknowability is what I would name the unconscious. On the level of the psyche, we can never consciously grasp ourselves once and for all because we are brought into being only within this Imaginary. As we rework the material of the Imaginary, we also draw boundaries between the conscious and the unconscious, the Imaginary and the Symbolic. But these boundaries are just that: drawn. Even if these boundaries have an imaginary dimension, we need to be able to project ourselves as free. This projection, which I identify with psychic separation, is what allows us to claim our person. This projection, in turn, allows us to distinguish between entrapment in an imaginary configuration and a claim to our own productive imagination. Dignity is the moral law that demands we mark our individuation from others as we see fit at any particular time in our life's journey. On the other side of this mark or line is an imagination that is our own, so much so that we can postulate the difference between the productive or radical imagination directed toward the future and an Imaginary that captures us in the repetition of the past. The distinction between Cardinal the writer and Cardinal the "mad" woman reminds us of the ethical importance of making the distinction between the Imaginary and the productive or radical imagination. I disagree with Castoriadis that the

Imaginary should be identified with the radical imagination. Yet I agree with him that the Imaginary remains in constant play in the reworking of the subject of desire.[64] For, in the case of Cardinal, she reimagines herself through the hypothetical fantasies she projected as the story of herself in the course of her analysis.

The psychoanalytically informed interpretation of dignity frees this ideal from its overly rationalistic and individualist roots. Indeed, it is the radical nature of the inherent critique of individualism as the truth of the subject that gives dignity a new urgency. My argument is not just that we are *socially* constructed. It is that we are intergenerationally produced through the law that *in principle* shapes us as subjects who can claim their own desire. That we are born as vulnerable creatures completely dependent on the good will of Others is what makes us fragile. These Others are not simply individuals. They are people shaped by the conventions of their times. They are inevitably traversed by their own primary Others, which is also why it is possible for their conscious and unconscious "secrets" to be passed on to the following generations. In this sense, phantoms are only too real. But their history is never simply private since they, too, are formed by history, culture, and society. Cardinal's mother could not pass on the gestures of rebellion against the restrictions of patriarchal culture. She did not know what those gestures might be. Despair for her was the condition of femininity, and *that* femininity was in turn completely intertwined with the conventions of the colonial lady. After her mother's death, Cardinal became a feminist because she realized that her mother's dilemma, and with it the cruelty to her daughter, was not hers alone. Such a dilemma could not simply be solved privately. It was completely intertwined with culturally and legally enforced norms of femininity, which denied the possibility of claiming dignity and freedom for women; even for upper-middle-class white women.

Dignity precedes the subject because without its recognition we can be forever caught in the *jouissance* of the big Other. Dignity is the name for the law that must be transmitted across generations. Since *jouissance* is a legal term that means the use of another subject without having to justify a claim on her, dignity is the barrier to such use, installed both in the name of desire and in the name of reason.

Following Martin Heidegger's reading of Kant,[65] I have given an existentialist-inspired dimension to the agency of our value-conferring capacity whenever we judge an object of our desire also as an end. Heidegger reminds us that, because we are thrown into a world that presents us with situations that call for an ethical or moral response, we cannot help making judgments and evaluations about what we say and do. As part of our moral awakening that we *do* make judgments and evaluations, moral freedom is a practice of self-responsibility we *must* assume. When we make these moral judgments and evaluations, we define who we are morally. We exercise our freedom as a narration of our self-responsibility that renders the value-conferring moment, in our actions and judgments, intelligible as our being called upon to justify ourselves to others with reason and rationality.[66]

Psychoanalysis deepens our understanding of this notion of moral self-responsibility in at least two ways. We can no longer simply presume the rational subject as capable of pursuing his or her desire. The actual existence of the desiring subject is simply taken for granted in much political philosophy. But psychoanalysis helps us avoid the error of both subjective idealism and positivistic psychology, neither of which can explain the *desire* to reconcile one's own freedom with that of others. Since we are dependent on others, and we come into existence as subjects through the discourse of the big Other, our freedom is always social and relational. There is no better example of this dependence than that of the relationship between mothers and daughters. An "unfree" mother, cut off from her own desire and, correspondingly, from the knowledge of her responsibility for her own life cannot pass on the value of freedom to her daughter. Castoriadis explains how psychoanalysis helps us understand that *in principle* we can desire the freedom of the other:

> In reality, however, it is because the autonomy of the other is not absolute fulguration and sheer spontaneity that I can aim at its development. It is because autonomy is not the pure and simple elimination of the discourse of the other but the elaboration of this discourse, in which the other is not an indifferent material but counts for the content of what is said, that an intersubjective action is actually possible and that it is not condemned to remain useless or to violate by its existence what it posits as its principle. It is for this reason that there can be a politics of freedom and that we are not reduced to choosing between silence and manipulation, consoling ourselves with "after all, the other will do whatever he wants with it." It is for this reason that I am finally responsible for what I say (and for what I leave unsaid).
>
> The final reason for beginning with the autonomy of the individual is because autonomy, as we have defined it, leads directly to the political and social problem. The conception we have discussed shows both that one cannot want autonomy without wanting it for everyone and that its realization cannot be conceived of in its full scope except as a collective enterprise.[67]

Feminism is such a collective enterprise. Unless someone is able to claim desire as her own, she does not even know what she is reasoning about. In Cardinal's case, her capacity to reason was completely shut down by what she described as her living nightmare. She had to find herself before she could begin to think, let alone write.

We can now begin to see why psychoanalysis is not something that is exclusively private. But this is not simply because of the symbolic and thus public nature of both sexual difference and gender roles. The understanding of the unconscious as the discourse of the Other—the Other who speaks through and commands the subject—makes the question of the survival of the subject at once ethical and political.

Feminism, Ego Ideals, and the Person

If dignity can never be lost because it is always ours *in principle*, then it also needs to be claimed. As I have described it, the law of dignity is structurally similar to the law of symbolic castration. In Lacan, symbolic castration is a law of psychic separation that the infant must inevitably confront because the mother or primary Other is a separate being who is, *in principle*, free. In her path-breaking work, Jessica Benjamin, an intersubjective analyst, has emphasized the importance of the mother claiming herself and her desire in the process of interrupting the traditional Oedipal pattern described in Anglo-American ego psychology, in which both sexes turn to the father to find freedom from the mother-child dyad. In *The Bonds of Love*, Benjamin argues that feminists, as well as psychoanalytic theorists and clinicians, have failed to come to terms with the crucial place of desire in women's subjectivity.[68] Within feminism this turning away from women's desire led some aspects of the movement to depart from freedom and move toward Puritanism. I have called this kind of feminism "white-knuckling feminism."[69]

For Benjamin, rigid gender identities in which we are either one or the other are endemic to the Oedipal constellation. But this constellation is not the end of the story. By focusing on the mother's activity in child rearing, which includes states of being with the child—wrongly described as passive, according to Benjamin—we can begin to understand aspects of the pre-Oedipal identification with the mother as part of a more complex, inclusive, and rich set of identifications. In this way, she gives an affirmative dimension to the Imaginary by insisting that our early identifications are not simply negative. They are formative of ego ideals, which play a crucial role in the psychic life of an adult human being.

As a clinician who has worked in a patriarchal culture, Benjamin realizes that the Oedipal constellation, as traditionally interpreted in ego psychology, continues to retain its psychic hold in the lives of actual men and women. Benjamin describes the result of this constellation for femininity:

> This constellation, I propose, is actually the boy's oedipal posture in which the identification with the mother is repudiated, and the elements of passivity associated with his babyhood are projected onto the girl, the daughter. This projection determines the daughter's position as that of a passive container for the male's defensively organized activity. Accepting this feminine passivity in her oedipal turn toward the father, the daughter's position, upon which Freud and his followers so vehemently insisted, does appear to constitute the content of femininity as we know it. Activity and passivity are thus divided in the oedipal position, foreclosing the possibility of being a subject who would own both tendencies within the self.[70]

To overcome the dichotomy of passivity and activity, which circumscribes the masculine and the feminine positions, Benjamin argues that activity includes capacities such as patience for, and ownership of, our desire. Benjamin seeks to return the mother to the family picture, but not just by reevaluating aspects of the received wisdom regarding the maternal relationship to the child. A woman who can own her desire is also available as a continuing object of identification since, as a desiring subject, she too can represent freedom to the child. Benjamin emphasizes negativity in the maternal relationship between mother and child, and any other primary Others, as crucial to psychic birth. The mother is never simply the one who holds and helps the child contain stimulation; she is also the one who struggles with the child and survives the struggle as another self. If the mother has relinquished her claim to her own subjectivity, then, from the outset, there is no self to engage with the child.

For a Lacanian analyst, the mother who survives is the symbolic mother, the one who can enter into a meaningful relationship with her child. Benjamin does not make the distinction between the imaginary and the symbolic mother. As we have seen in my discussion of Gurewich, the mother of the Imaginary is the object of the child's incestuous longing. It is this imaginary figure to whom the child must say goodbye if she is to achieve a life of her own. But that does not mean there is nothing left of this early relationship in

terms of the ego ideals that have been formed through it. Benjamin emphasizes the ideal aspect of these ego ideals, which actually involve the symbolization of the relationship. I agree with her that ego ideals allow women as well
as men to transcend rigid Oedipal complementarity so that they can tolerate
ambivalence—the back-and-forth movement made possible by fluid identifications—and muster the courage to face up to the violence and aggression inseparable from the struggle to "be a self" in the first place.[71] However, rather
than refer to the ideal as the *self*, I return to the term *person* precisely to mark
this ideal. I do this for two main reasons.

First, we should turn to the level of the *psyche*, which proceeds through
identifications out of which ego ideals are shaped. The use of the word *person*
evokes the constant movement that is so crucial for keeping alive the fluidity
and the play of gender identifications that Benjamin at once hopes for and
advocates. What we think of as traditional gender identifications function on
this level. They are formed through our primordial identifications with Others who, through words and actions, give us ideals by which our selves are
shaped as men and women. These ideals often contain normative injunctions:
"Men are macho like me," "Women must care about being thin," and so on.

But feminism is also an ego ideal. Although ego ideals are inevitably
shaped by our identifications with primary Others, they are never simply appropriated. On the level of the psyche, "I am a feminist," can mean many
things because every individual woman will give her own shape to this ideal as
she embraces it as an identification all her "own."[72] All the women in my family
are feminists. My sister, Jill, is a feminist. Her fourteen-year-old daughter,
Dylan, is a feminist. My nine-year-old daughter, Sarita, when she was explicitly
asked that question, answered, "There is no other way for me to be." And I am,
of course, a feminist. Indeed, I agree with my daughter that there is no other
way for me "to be." For my daughter, the meaning of her feminism will obviously be different than my own because she is a Latina and a dual citizen of
Paraguay and the United States, whereas I am a white Anglo and a citizen of
the United States. We have yet to see what she will make of her own feminism.

What feminism has meant for each one of the women in my family is different. For my sister, who is now president of an advertising company, Raxau
(which in Catalan means "a little bit of redeemed madness"), feminism has
meant building a "woman's culture" at work. Since she has not been em-
broiled in the battles over essentialism, she has not worried much about
whether her feminism is tainted by it. What a "woman's culture" means is a
place where women can flourish. This gets translated into a series of policies

for her company concerning flexible hours, maternity and paternity leave, and sexual harassment, among other considerations. But it also means a conscious effort to promote women so that they can realize their capacities. For my niece, feminism is translated into the way she relates to boys, her developing sexuality, and her aspirations for her life. But feminism can be an identification that works back in time. Before dying, my grandmother described her life as feminist, and this identification allowed her to do what she had not done for most of her life: articulate the reasons for many of her decisions, like running the company herself and refusing to marry again. Eva Perón was very much a part of my grandmother's feminism. Given my grandmother's profound identification with her, how could she not have been? The debate over whether Eva Perón was a feminist, or even had feminist impulses that were sincere, may go on for an eternity. But in my family she was so imagined.

In her exercise of her right to die, my mother described herself as dying like a feminist. What would it have meant both for me and for my mother to tell her she was not dying as a feminist because her last action was not in accordance with some pre-given definition of feminism? Certainly, it would not have been respectful of her dignity. Feminism, understood psychically as an ego ideal, tempers any attempt to create a feminist dogmatism. One crucial aspect of feminism's power is that it opens up new identifications coherent with the word woman and, correspondingly, with the word *man*. It challenges rigid gender identity. Benjamin's ideal of the self is clearly informed by exactly the spirit of generosity that stems from recognizing that feminism does not just present us with ethical and political ideals, but itself serves as an ego ideal that shapes the person.

As I use it, the term *person* can clearly encompass the play of *personae* that I claim inheres in the recognition of feminism as an ego ideal. But it is also consistent with Benjamin's emphasis on ownership and authorship of women's desire. For Benjamin, this idea of ownership is meant to challenge the Oedipal division of the genders into passive and active:

> Here I propose that authorship, or ownership, of our desire and intention
> is a crucial feature of subjectivity occluded by the conventional opposition
> between activity and passivity. As in my earlier discussion of "a desire of
> one's own," I show that being a subject of desire requires ownership and
> not merely activity. Ownership depends upon reclaiming the maternal form
> of activity, the recognition and holding of emotional states, excitement in
> particular. While we use the patrocentric theory to understand the form

conventionally assumed by femininity, we use the matrocentric theory to formulate the mother's contribution to our subjectivity and desire.[73]

For the moment, I want to leave aside Benjamin's concept of ownership, which involves a reevaluation of the role of the mother. I want instead to emphasize that ownership carries with it a concept of entitlement. This entitlement to claim one's desire and one's person is exactly what is usually legally and culturally denied to women in patriarchal society. The right to claim one's person stems from the dignity of the subject, which, as I have stressed, can never be taken away from any of us precisely because it exists only *in principle*. But we can certainly be denied the right to claim our desire and our person by an established legal system.[74] Thus, because the word *person* has legal connotations, I think it captures more effectively Benjamin's ideal, which demands fluid identifications as well as the right to own them in the form of a right to claim one's own person. On the basic level of self-image, this ideal of the person activates the dignity of the subject.

Because of the legal custom of referring to the subject of rights as a person, this notion can be easily translated into a modern Western legal system. Here we can begin to see how psychoanalytically informed feminism may influence feminist programs of legal reform. Unlike the Lacanians who battled against gay and lesbian rights, I do not believe there is or should be any simple and direct connection between legal reform and psychoanalytically inspired conceptions of the subject. Like the philosopher John Rawls, I believe that we should propose ideals for legal reform that could be agreed upon by those who hold many different positions on such basic philosophical ideas as the subject.[75] In the international arena, and particularly when it comes to questions of human rights, I would insist that to demand such agreement inevitably obliges us to engage the worst aspects of our imperialist legacy.[76] Nevertheless, this ideal of the person can inform our attempts at legal and political reform in the United States and Western Europe. For such attempts would be tempered by the recognition of how important it is for feminism to retain its power as an ego ideal, shaped differently by different women. A feminism that insists on the psychic entitlement as well as the legal right of women to claim their own person—and, with their person, their desire—will be reluctant to label as false consciousness women's attempts to pursue this goal. If as a person she is placed at the level of the ego and the Imaginary, then she would always be reinventing and transforming what it means for her to claim her desire. There is no such thing as an authentic desire that is absolutely true to feminist aspirations. There is no bright line between good desires and bad desires because the whole point is to emphasize the importance of women claiming their desiring subjectivity. In the name of dignity, and its translation into the ideal of the person, we should keep the psychic space open for women to begin to act out their desires, to see what happens and how they will change. In this way, my ideal of the person and Benjamin's ideal of the self come together. They are both meant to emphasize the value of freedom and, in particular, a woman's entitlement to her desire.

Some Thoughts on the Word Care

It is obvious that Cardinal's mother could not imagine her own dignity and therefore claim her own person. Since she could not claim it, she could not transmit her sense of dignity, much less a woman's entitlement to her person and desire, to her daughter. In the care/justice debate, the two are often opposed to one another or are said to serve different registers. But under the view of psychoanalysis I have offered, the greatest gift a mother can give to her daughter is the insistence on the daughter's freedom, desire, dignity, and entitlement to her own person. This is how my grandmother transmitted feminism to me, even before she had "the words to say it." Of course, children need what we think of as care. However, in her recent book, *All About Love: New Visions*, bell hooks writes that she was always cared for but never loved. She means that there was always food on the table, clothes to wear, someone to pick her up at school, but she never felt seen, and since she was not seen, her aspirations could not be respected.[77] Respect for the dignity of the Other, particularly of children, is crucial to love.

Sometimes we "care" for our children best by not catering to their needs. For a Lacanian, this kind of care actually can be disastrous because the primary Other is confusing need with demand. Sometimes, when the baby cries, she wants milk; sometimes she wants reassurance. She wants to know that "everything is okay." Later on in life, we are all too often met with the command to go away and sometimes that is exactly what we should do. Certainly, any of the sophisticated writers on *care* would include the complexity of actual mother-and-daughter relationships in the word. My worry about the word, then, is that it is too readily associated with passivity and not with the struggle between mother and daughter that will inevitably ensue, if both insist on their dignity. To clarify just how central dignity is in intergenerational relationships, I would use the expression *affective and ethical attunement* in the place of *care*. Since attention to our daughters' freedom and dignity, as well as to our own, will bring us into conflict with patriarchal assumptions, the endurance this kind of attunement demands may well psychically fuel a sense of justice. In my own case, my daughter's Latin American heritage has not only strengthened anti-racist politics in my life, but has also inspired me to rethink many of my basic political assumptions about language, citizenship, and ethnic as well as racial identity.

What Remains of Sexual Difference?

Some Lacanians would strongly object that I have neutralized the law of the paternal function, and indeed I have, if by "neutralized" one means that I have continued to use a "scouring pad" against Lacan's patriarchal assumptions. This is not only because there is an ideal aspect to dignity—which I believe actually captures the ethical goal of psychoanalysis, at least as defended by Gurewich—but also because I do not believe that the basic psychic law of separation needs to be named as masculine. In her insightful work, Elizabeth Bronfen refers to this law as representable through the scar of the navel. She refers to the process of coming to terms with the finitude and separateness that the scar represents as "denaveling."[78] By denaveling, Bronfen means the process by which we all learn to accept that we are born to die and that there is no going back to a oneness with the primary Other that would save us from this fate. Obviously, the navel is something both sexes share. But there is no structural reason for the law of psychic separation, as Gurewich understands it, to be identified as masculine. If we must make reference to body parts, why not represent that separation as the navel? As the symbol of that law that gives us to live and die at the same time?

In his last seminars, Lacan clearly separated the conflation of the symbolic father with the actual father. The names of the father are who or what the mother desires. So why not use the abstract name *dignity* for those names? If, as in Gurewich, the Oedipal complex is an effect and not a cause of subjectivity, it becomes more difficult to make the argument that the masculinization of desire is a meta-psychological consequence of psychic separation. At the end of Lacan's life, the idea that desire in both sexes is masculine is the only justification for the position that the feminine within sexual difference cannot be represented and that desire cannot, as a *structural matter*, be expressed as feminine.[79] Desire, in other words, need not be phallic. Gurewich is one of the few thinkers to discuss what happens to a woman psychically if she resolves her Oedipal complex. A woman who accepts the necessity of psychic separation can claim her desire as her own and thus name it "feminine." To explain why desire is not necessarily phallic in Lacan's theory, we need to return to Gurewich.[80]

Gurewich challenges the idea that the *jouissance* of women and, by extension, women's desire, is fated to be the mysterious Beyond of the social and symbolic contract. She does this by arguing that Lacan replaced the mandate of sexual difference with the law of symbolic castration. I have in turn renamed that law of psychic separation *dignity*. Once these steps are taken, there is no longer a defensible philosophical reason for the unrepresentability of feminine desire. What remains of the feminine, if it is no longer placed outside the symbolic and social contract? My answer is that "what remains," after the destructive effects of the Oedipal fantasy have been dispelled, is the capacity to express the *feminine within the imaginary domain*. The feminine within the imaginary domain and, with it, symbolizations of feminine sexual difference are endlessly brought into being, always with the future as their point of reference. There is no end to "it," in a positive cognizable reality, no end to what women dare to be.

Let me clarify, then, what I mean today by the *feminine within the imaginary domain*. I do not mean that it is an attribute of woman; nor that "it" is in any way associated with femininity. Once we have dispelled the fantasy of the Oedipal myth as traditional psychoanalysis has perceived it, the feminine within the imaginary domain is the way we might reimagine women and redefine them. Integral to this endless reimagining and redefining of the meaning of women is the recognition that we can be women differently, since "it" is in the imagination and the languages in which we give "it" meaning. "It" does not exist in itself, but only as "it" is expressed and articulated. Because *no one is just a woman*, our sexual difference is always coming into being as actual women endlessly live out different *personae* for themselves. As a Latina, my daughter is identified on the side of the feminine by the Spanish language, but that does not mean that her sexual difference can be expressed separately from her Latin heritage. In other words, I mean to replace both the idea of "gender and ... *something-or-other*" and the notion of "intersectionality." Intersectionality attempted to show that gender could be defined only within an intersection of race, nationality, and linguistic heritage. The point is well taken, but I wish to make it stronger. Latina is an *ethical* affirmation that the Spanish language, ethnic origin, and an avowed family connection to South America, Puerto Rico, Cuba, or Mexico, as opposed to Spain, is *integral* to the way in which the feminine within the imaginary domain is envisioned and rearticulated. In this way, feminism is by definition multicultural and committed to transnational literacy. Without this commitment, we cannot hope to engage the ever-changing meanings of the feminine within the imaginary domain. The definition of Latina is always changing as it is lived out in the lives of those who identify as Latinas. As I am using the phrase *the feminine within the imaginary domain*, it serves freedom and the place of the aesthetic I have made central to feminist theory. Since it insists on the *might yet be*, the feminine within the imaginary domain also operates against the reification of any given symbolizations of Woman or women. When there is always more to come, there can never be a last word. There is only time to accept the ultimate daring of the feminine, time to look our mothers in the eye and hear their words of freedom resound, resound, resound . . . "From now on, call me by my *own* name."

Chapter III

THE ART OF WITNESSING AND THE COMMUNITY OF THE *OUGHT TO BE*

*In the chain and continuum, I am but one link. The story is me, nei-
ther me nor mine. It does not really belong to me, and while I feel
greatly responsible for it, I also enjoy the irresponsibility of the pleasure obtained through the process of transferring. Pleasure in the
copy, pleasure in the reproduction. No repetition can ever be identical,
but my story carries with it their stories, their history, and our story
repeats itself endlessly despite our persistence in denying it. I don't
believe it. That story could not happen today. Then someday our
children will speak about us present, about those days when things
like that could happen.*[1]

—*Trinh T. Minh-ha, "Grandma's Story"*

REMEMBERING HER RIGHT

Trinh T. Minh-ha's "Grandma's Story" evokes the power of women's stories to
connect generations to generations and keep alive a history that would otherwise
die out. From these stories and their power, the woman writer shapes herself.
This oral tradition keeps "alive the forgotten, dead-ended, turned-into-stone
parts of ourselves."[2]

When there are stories to remember and pass down, there can be witnesses.
As brutal as it is, a story that can be told can be engaged, accepted, or recreated.
But what if there are no records and no stories because they have been buried
under the legacy of imperialism? What if, when one looks for the story, only the
story that the colonizer tells about "them" is found? How does a writer present

such silence in order for it to echo? What, then, does it mean to witness? These are just some of the questions that Gayatri Spivak addresses in her extraordinary chapter on history in her book *The Critique of Postcolonial Reason*.[3] Spivak here returns to the question she asked some years ago in her rightfully well-known essay "Can the Subaltern Speak?"[4]

Spivak proceeds to deepen our understanding of feminist witnessing through her tracking of the figure of the "native informant."[5] The native informant is not a figure in the traditional sense because she is a "trace," and her story exists only as "a subliminal and discontinuous emergence."[6] Spivak uses psychoanalysis to explain how unlisted traces can remain for us as what is missing as we begin to measure silence and pay heed to the significance in the cuts of the threads that should tie together a literary text or a historical narrative.[7] Spivak actually addresses two kinds of silences. The first kind of silence she represents is her own failure to recover effectively the history of the Rani of Sirmur who threatened to commit *Sati*, but did not. *Sati* is a form of ritual suicide practiced by some widows in India who choose to burn with their husband after his death. The other kind of silence Spivak seeks to represent is not silence as such but rather a misinterpretation based on stereotypical understandings of women's actions. Spivak tells us of a distant relative who desperately strove to give notice of her ethical and political dilemma through the drastic action of suicide. As a national liberation fighter, she was ordered to commit a political assassination and realized she could not do it. Her suicide was completely misunderstood by the very people to whom she wanted to communicate the unbearable conflict that made it impossible for her to continue living: her family and friends. They thought she had put an end to her life because of an illicit love affair. What Spivak seeks to uncover is the motivation of Bhubaneswari's decision to hang herself. It was the failure of her own relatives to "listen" to the significance of Bhubaneswari's suicide that led Spivak to question despairingly whether the gendered subaltern could ever be heard.[8]

But now, let us turn to the "stories." I put "stories" in quotation marks here because Spivak actually writes about what cannot be effectively narrated. In the case of the Rani of Sirmur, her story is about why her history cannot be written. Spivak's representation of the misinterpretation of Bhubaneswari's suicide is an account of how such a dramatic action was trivialized by understanding it as one more story about a woman making the foolish "female mistake" of getting pregnant outside of wedlock.

Spivak did not initially intend to track this particular Rani, the Rani of Sirmur. She was more generally interested in *Sati* and women who practice *Sati* as ritual mourning. *Sati* became a British obsession when other ritually accepted forms of suicides were not subjected to their regulation. As Spivak explains:

> The first category of sanctioned suicides arises out of *tatvajnana*, or the knowledge of right principles. Here the knowing subject comprehends the insubstantiality or mere phenomenality (which may be the same thing as nonphenomenality) of its identity. At a certain point in time, *tat tva* was interpreted as "that you," but even without that, *tatva* is thatness or quiddity. Thus, this enlightened self truly knows the "that"-ness of its identity. Its demolition of that identity is not *atmaghata* (a killing of the self). The paradox of knowing the limits of knowledge is that the strongest assertion of agency, to negate the possibility of agency, cannot be an example of itself.[9]

These religious suicides were actually considered non-suicides. *Sati*, on the other hand, was not prescribed as an exception to the rule against suicide in the Hindu religion. So women who practiced *Sati* were *not* following the prescriptions of their religion. There was no religious pressure put on them to undertake this form of ritual mourning. Nor was there any shame in a woman changing her mind at the last moment. However, the British created a kind of "nationalist shame" for those women who did not commit *Sati*. Only after the British had intervened by reading a statement advising women to think twice before every practice of *Sati* did a woman, who turned back on her decision, undergo a particular kind of penance. The only religious "advantage" of *Sati* was the belief that, by burning now, a woman could avoid being born in a female body again. However, *Sati* was not, despite all the myths about it, in the general customary code of a widow's behavior.

Who were these women, then, who exceeded custom and obligatory religious behavior? And why did "saving" them come to play such an important role as key evidence of the British "civilizing mission" in India? We can begin to see why Spivak turns to *Sati*, and the reactions to it as a possible example, to help demonstrate the ethical and political power of the central thesis she develops throughout her book. That thesis is summarized by Spivak as follows: "'Germany' may have taught us to think the ethical subject, but Imperialism used Woman 'freeing' her to legitimize itself."[10]

Spivak's intuition is that "the ground level value-codings that write these women's lives elude us."[11] But of course she wants to test whether she is right about that. In her preliminary challenge to the received story of women who practiced *Sati*, Spivak refuses them the pity she believes has been evoked by

the received Western version of their story, or by the sentimental, purportedly nationalistic response to it:

> The Hindu widow ascends the pyre of the dead husband and immolates her-
> self upon it. This is widow sacrifice. (The conventional transcription of the
> Sanskrit word for the widow would be *sati*. The early colonial British tran-
> scribed it *suttee*). The rite was not practiced universally and was not caste- or
> class-fixed. The abolition of this rite by the British has been generally un-
> derstood as a case of "White men saving brown women from brown men."
> White women—from the nineteenth-century British Missionary Registers
> to Mary Daly—have not produced an alternative understanding. Against
> this is the Indian nativist statement, a parody of the nostalgia for lost ori-
> gins: "The women wanted to die," still being advanced.
>
> Faced with the dialectically interlocking sentences that are constructible
> as "White men are saving brown women from brown men" and "The women
> wanted to die," the metropolitan feminist migrant (removed from the actual
> theater of decolonization) asks the question of simple semiosis—What does
> this signify?—and begins to plot a history.[12]

Through this gesture alone, she challenges the two social imaginaries that
clash in the received story. I am using Cornelius Castoriadis's term for a "social
imaginary."[13] A social imaginary implies not only that there are collective signifiers—the signs we use to give meaning to our world—but also that there are
signifieds—the objects we perceive as simply there in reality—that are irreducible to those perceived or narrated by an individual person. A social imaginary often produces a conflation of the two in such a way that what we envision
as reality loses its status as a vision or perception and simply seems foundational of the world as it is. "Native woman who commits *Sati*" is a signifier that
is perceived differently, as signified, by the British. A social imaginary describes
those parts of our social world upon which we are so dependent for our orientation to our reality. We can barely see them and therefore cannot imagine
them being changed. They simply appear as reality, as the backdrop of our conscious thoughts, including our reasoning process. Spivak quotes from Edward
Thompson's book *Sutee* to give us a sense of how the British colonial imaginary
saw the problem of *Sati* and the women who practiced it:

> Consider in that frame Edward Thompson's words of praise for General
> Charles Hervey's appreciation of the problem of *Sati*: "Hervey has a passage
> which brings out the pity of a system which looked only for prettiness and

constancy in woman. He obtained the names of the satis who had died on the pyres of Bikanir Rajas; they were such names as 'Ray Queen, Sun-Ray, Love's Delight, Garland, Virtue Found, Echo, Soft Eye, Comfort, Moonbeam, Love-lorn, Dear Heart, Eye-play, Arbour-born, Smile, Love-bud, Glad Omen, Mist-clad or Cloud-sprung—the last a favourite name.'"[14]

These "olympian violations of women's names,"[15] to use Spivak's phrase, clearly lead to a dead end. Even in her most imaginative efforts, Spivak could not begin to translate this list, which she contrasts with the detailed information available on most of the British cadets' reports in the East India Company. The problem was not with the British ability to keep records. Spivak finally does find a Rani who threatened *Sati* and made it into history. But what can Spivak find out about this woman? First, and much to her regret, Spivak cannot uncover her name: "We are not sure of her name. She is once referred to as Rani Gulani and once as Gulari. In general she is referred to, properly, as the Ranee by the higher officers of the Company, and 'this Ranny' by Geoffrey Birch and Robert Ross."[16] She remains only as the Rani of Sirmur. What does Spivak find out about her? Her husband the Raja of Sirmur, Karam Prakash, was deposed and banished from Sirmur by the British for his dissoluteness. But his was a frequent charge against many other chiefs, and nothing ever happened to them. As Spivak finds out, then, by looking at secret correspondence now available in the archives, dissoluteness alone was no reason for banishment: the only other reason was that he had syphilis, which was referred to as his "loathsome disease." The Rani was then established as the official guardian of the still-minor king, her son Fatteh Prakhash. Supposedly, there were no adult responsible male relatives who could take charge after the Rajah's banishment. But Captain Geoffrey Birch actually praised one such male relative, Duleep Singh, for his "astuteness." In fact, the absence of a senior male relative seemed an unlikely reason for not replacing the banished Rajah. The records of the East India Company show that the company had plans for the dismemberment of Sirmur. By annexing the entire half of Sirmur, the company could protect its routes and frontier against Nepal. The Rani and her minor son were not in much of a position to stir up a fuss about that, so the British hoped to force a settlement through her. Spivak has no trouble locating the names of all the men in this drama. The Rani of Sirmur was given executive authority over her son and, through him, over her subjects, but of course under the watchful eyes of Captain Birch.

Two pieces of information appear about her in Birch's record: two of her husband's ex-wives who had been separated from the household for fear of intrigues asked to come back and she took them in. The Rani also managed to reinstate a pension for a great-aunt who had been cut off after a quarrel with her husband. She negotiated a deal with her great-aunt beginning with 700 rupees—although she had actually put away 900 rupees for this purpose, expecting her great aunt to ask for more. As Sir David Ochterlony wrote: "It has been necessary for Captain Birch occasionally to interfere with her authoritatively to counteract the facility of the Ranee's disposition."[17]

It was then that she announced her intention to commit *Sati*, which was extremely inconvenient for the British. She could not be offered the free choice to commit *Sati*, since that would get in the way of the settlement the British wanted her to accomplish. As a result, Birch asked for governmental authority to intervene and keep her from acting on her decision. He specifically asked for a government regulation. The governor's secretary's letter seems to refuse Birch direct authority to stop the Rani from committing *Sati*, but suggested alternative routes, such as the appeal to religious authority to convince her to do otherwise. It was a point in history in which the practice of *Sati* had not yet been abolished, and a time when the British still respected "the religious prejudice of natives amongst its own subjects." The *pundits* and the *brahmans*, it is clear from the letter, were to be manipulated to give her the advice: she could not even perform a legal *Sati* because her interests were now formally separated form her husband's.

In fact, as Spivak points out, it was the British government that had so separated her interests from those of her husband. Here we find the real incentive: credit was extended only to wives who were alone. The government further authorized Birch both to continue to keep her husband's messages from her and to communicate to him that he was not to contact her. Both were to be told that her paramount duties to her son and to her subjects superseded any she might have to her husband. "Paramount," Spivak notes, is the word used for imperial power.[18] In reality, then, the government washed its hands by giving Birch the authority to use every means of influence and persuasion to prevent the Rani's *Sati*. Birch interpreted the Rani as wanting to escape from her predicament, and to be reunited with her husband. After further consultation, the Rajah's banishment to a farther part of India was postponed. And then the Rajah's fate becomes unclear and the matter vanishes from the British correspondence—but not from Spivak's continuous search for the answer that will uncover whether Birch's interpretation of the Rani's desire was right.

Spivak sets out on a trail to find out what happened to the Rani, and she searches for those clues that might signal what her desire was at the moment when she announced her decision to commit *Sati*. Spivak goes to Nahan, the county seat in the modern district of Sirmur. The records she finds there begin with the reign of the Fateh Prakash, the Rani's son, now presented as similar to the European model of "native kings." In order to get into the palace, which is occupied by an elected representative of the state assembly, Spivak has to embroil herself in litigation between Fateh Prakash's descendants—who ended in a female queen of Jaipur at the time of Spivak's research—and the male descendant from the younger male line, who was occupying the vanished throne. The original royal line of Sirmur no longer existed. Even so, the seat was being contested by the Maharani of Jaipuir. It is under the auspices of the Jaipur's representative that Spivak is finally allowed into the castle. She spends the night there. Like the Rani of Sirmur, she is locked in; no outsider is allowed to be there since such a presence might influence the litigation. She does not even have access to the female quarters, which she presumes were the Rani's. Instead, she is put in the men's quarters, which were added on in the nineteenth century. She searches for any signs of the Rani but finds none. All in all, Spivak makes five trips to the palace. We can imagine her being locked in that palace trying to find some indications of the Rani's rattling around. Spivak asks her readers to imagine the other scene of a locked-in woman:

> We imagine her in her crumbling palace, separated from the authority of her no doubt patriarchal and dissolute husband, suddenly managed by a young white man in her own household. Such examples must be accommodated within the epistemic violence of the worlding of worlds that I have described above. For this too is the sudden appearance of an alien agent of "true" history in native space. There is no romance to be found here. Caught thus between patriarchy and imperialism, she is a representative predicament, a woman whose "exchange," from "feudal" to "modern," as the agent of her subject-child, will establish historicity.[19]

Spivak cannot even know enough of the Rani's own voice to engage in the transferential relationship the historian Dominick LaCapra suggests is a useful fiction in archival work.[20] Using the psychoanalytic metaphor of transference, La Capra's suggestion is that the archivist can expand the imagination and hear voices from another imaginary by engaging with the original text as if those voices could talk back to us. This fiction is to help the archivist protect himself against projection of meaning onto someone from another culture and from another time. But as Spivak tenderly notes, she does not even have enough of the Rani's voice to imagine her talking back:

> I should have liked to establish a transferential relationship with the Rani of Sirmur. I pray instead to be haunted by her slight ghost, bypassing the arrogance of the cure. There is not much text in her name in the archives. And of course there is no pretense of continuity of cultural inscription between her soul and the mental theater of the archivists. To establish something like a simulacrum of continuity is that "epistemic violation" that I invoked in my more turgid phase. It started in the Rani's son's generation. She was only the instrumental agent of the settlement.[21]

Spivak does, however, find out the end of the story, at least in the literal sense. She finds an account of the Rani's death in an alternative record. She died in 1837, and the report of her funeral makes it clear that she did not commit *Sati*. Why she did not, we will never know; it escapes us. On her last visit, Spivak remembers her attempt to find the Rani's palace, knowing now that she will never be coming back and that there is no place else to go to find the Rani: "On the very first try, in search of the palace, I had walked about in the hills where buses did not go. Shy hardy women gathered leaves and vegetation from the hillside to feed their goats. They could not have had a historical memory of the Rani. And they are, have been historically, at a distance from the culture of imperialism, and from the relay between princely state and nation-state that swept the Rani's descendants into its currents and whirlpools."[22]

Does Spivak have a position on widow burning? Of course she does. She is against it. But she wants to change the scene of the ideological battleground to make room for the possibility that the women who practiced *Sati* cannot be reduced simply to victims. Certainly she knows that she is staging the battleground, but the ethical purpose of that restaging is precisely to preserve the *pathos* of the burnt widows. If we strip them of their dignity, at least as we imagine it might have been exercised then, their deaths lose their *pathos*. We deny them the significance of the agency they might have exercised. And what is Spivak's counternarrative?:

> I will work toward the conclusion that widow-sacrifice was a manipulation of female *subject*-formation by way of a constructed counter-narrative of woman's consciousness, thus woman's being, thus woman's being-good, thus

the good woman's desire, thus woman's desire; so that, since *Sati* was not the invariable rule for widows, this sanctioned suicide could paradoxically become the signifer of woman as exception. I will suggest that the British ignore the space of *Sati* as an ideological battleground, and construct the woman as an *object* of slaughter, the saving of which can mark the moment when not only a civil but a good society is born out off domestic chaos. Between patriarchal subject-formation and imperialist object-constitution, it is the place of the free will or agency of the sexed subject as female that is successfully effaced.[23]

To construct this narrative, Spivak begins a journey in search of the Rani, a journey that becomes a story of fadeouts. Spivak sadly notes that the Rani was entered into history, but her death and life could not be commemorated. By miming her journey and demanding that we imagine ourselves with her, she refuses to let the *pathos* of what might have been the Rani's life fade out. She not only dramatically presents what cannot be smoothly narrated, but also asks us to join her in imagining the significance of what imperialist history has done to the Rani of Sirmur:

> The narrative pathos . . . is at a great remove from the austere practice of critical philosophy. Yet, the differential contaminations of absolute alterity (even to utter the words is to differentiate them from some other thing, which should of course be impossible) that allows us to mime responsibility to the other, cannot allow this pathos merely to be faded out. As I approached her house after a long series of detective maneuvers, I was miming the route of an unknowing, a progressive différance, an "experience" of how I could not know her.[24]

Spivak's obsession is clearly not with *Sati* but with how female subjectivity is effaced, particularly when it is enacted in a manner traditionally seen as pitiful. But she is also working against the nostalgia that pretends there were good old days to return to or that indeed could even be recovered. By insisting on restaging ideological battlegrounds so that the *might have been* of woman's agency can be returned to the picture, she keeps these women from being given a false burial—from being buried under a legacy of pity and victimization. At the end of her chapter on history, Spivak turns to another suicide, that of a woman who clearly means to express herself through the taking of her own life. Her stories are about different burials of the meanings of women's actions. Bhubaneswari was trying to say something with her suicide, and not even her own family of "emancipated" granddaughters could "hear" her. At sixteen or seventeen, Bhubaneswari Bhaduri hangs herself.

Once again, Spivak gets on the trail of the meaning of Bhubaneswari's suicide, which confused her family. Bhubaneswari hanged herself while menstruating to show that her action was not due to an illicit passion resulting in an unwanted pregnancy. Spivak knew of the suicide through family connections. Her nieces told her the family's version of the story: she had died because of an illegitimate love affair. Other family members wondered and questioned her about her wanting to focus on this hapless sister while the two other sisters had led such successful lives. Indeed, it was her own interest in Bhubaneswari's failure to communicate with anyone around her that led Spivak to despair over the subaltern being completely silenced. It turns out that Bhubaneswari committed suicide because she had joined a nationalist group and was ordered to carry out a political assassination. She was caught between two ethical commands—the command to carry out a murder and moral qualms about killing—and, in this sense, her suicide is classically tragic. It is tragic because she chose what she saw as the only ethical way to resolve her dilemma. Once again, Spivak returns the *pathos* to a woman's death by respecting the dignity of her action.

The *pathos* inheres in the magnitude of the actual voices of the subaltern. Precisely because there is so little documentation of this history, the enormity of the loss cannot itself be directly represented. When documents can be found, they are almost always those of the colonizer and not of the gendered subaltern. We, the readers, are asked to imagine the significance of the loss since Spivak herself can only come close to figuring the character of the woman or women she sets out to recover in their networks of resistance. Indeed, at least in the case of the Rani of Sirmur, she represents herself in the failure of her efforts at recovery. How, then, are we to understand Spivak's insistence that we take notice of her own failure? She continually asserts that there is an ethical and feminist purpose in her questioning of historical knowledge. The question remains: what, exactly, is that purpose?

THE PRACTICAL SUBLIME

In my reading of her, Spivak dramatically represents the ethical significance of what has been lost, fundamentally misunderstood, in the history of the gendered subaltern. This loss can be summarized as the "valued codings" of the women's lives as they themselves tried to make sense of their responsibilities to themselves and to others. How are we to grasp the feminist significance of this effort at dramatic representation? I interpret Spivak's feminist-inspired historical project through a reading of Immanuel Kant's *The Critique of Judgement* and Friedrich Schiller's interpretation of Kant's idea of the sublime. But first, I must say that my own voice is bound up with that dramatic representation.

I am trying to witness to Spivak—to Spivak *as* another woman. What cannot be told with narrative coherence, what could not be heard or recorded in the history of her own family—I am trying to bear witness to this. What Spivak herself seeks to represent is the very problem of representation itself. As an effort to preserve faithfully the remains of the women who have been either lost in history or effectively silenced by a misread interpretation of their actions by colonial governments, Spivak's history of "points of fadeout" forces us to confront a central dilemma of feminist history:

> Paradoxically, the retrieval of the history of the margin can be a lesson not only to the writing of woman's history triumphant, but also to the writing of the most hegemonic historical accounts. This is not merely a rhetorical device until enough research has been done. Or, if so, this is the strongest sense of rhetoric, which works at the silences between bits of language to see what will work as meaning, to ward off a silence filled with nothing but noise. . . . It must be kept in mind that by the account of these fadeout points, I represent the rhetorical limits of logic, which in turn disclose, by cordoning off, the violent limits of rhetoricity. No one can "present" them, or to present (them) is to represent. It must also be kept in mind that, given a different structure of authority and policing of the sub-Himalayan rural areas of India in the late eighteenth and nineteenth centuries, in trying to locate the Rani we may be groping in the margins of official Western history, but we are not among marginal women in their context.[25]

Spivak is, of course, aware of all the criticisms of a representational consciousness that simply purports to mirror the world.[26] Representation is never simply passive. Either consciously or unconsciously, the writer is representing herself when she writes history, even when she describes herself as only recording the representations of the individuals or groups who have documented themselves, such as workers cooperatives. Without this multi-layered understanding of how representation operates in the attempt to write history, particularly of any group of women of a former colony, "the willed (auto)biography of the West still masquerades as disinterested history, even when the critic presumes to touch its unconscious."[27]

Given that she is not trained as a historian and frankly admits that she has no archival experience, what initiated Spivak's turn to history? Once again, Spivak attempts to locate herself and spell out how she is representing herself in this project. Even as she is aware of all the complexities of her motivations, she rejects the idea that her attempt to trace the Rani of Sirmur is due to her political identification as a South Asianist. Furthermore, she does not claim that the Indian case is representative of countries and cultures that have also been evoked as the "Other of Europe." But birth and knowledge of language do allow her to make headway in her search for voices of the gendered subaltern. In other contexts, she would not have this basic access to their experience, whether recorded or recounted. I do not have that access. But as the reader, I am nevertheless called by Spivak to witness. In her call to her readers, Spivak is seeking to bring into being a particular kind of community. That community can best be understood as Kant's *sensus communis aestheticus*, *communis* indicating a community that arises in the aesthetic judgment that we are before a sublime or beautiful object or person. Since her focus is loss and fundamental misinterpretation, along with the tragedy of both, Spivak is calling us to judge them as sublime.

As a dramatist, Schiller was interested in presenting the sublimity of human experience. Unlike Schiller, Kant almost always uses natural objects when he seeks to evoke the sublime. But it is not the fear that arises in us because of a thunderstorm or tornado that makes the object sublime. Indeed, to the degree that we, as sentient beings, cower before the might of nature, the more we collapse in terror as the fragile and physical creatures that we are, the less we feel the sublimity of what is before us. Someone in such a state would be incapable of making the aesthetic judgment that what stands before her is sublime.

So what, then, does Kant mean by an aesthetic judgment? Taking note of two common mistakes about the quality of the aesthetic and the meaning of a reflective judgment will allow me to clear the ground for a definition. First of all, it is a fundamental confusion to limit Kant's own analysis of aesthetic judgment to any particular object field—nature, or as some current interpretations suggest, the realm of art.[28] I quote Kant himself to emphasize that it is an error to limit the aesthetic quality to a specific realm or territory of objects: "That which is purely subjective in the representation of an Object i.e., what constitutes its reference to the Subject, not to the object is its aesthetic quality."[29]

Kant's entire purpose in *The Critique of Judgement* is to find the place in his critical philosophy for the subject and her feelings. But Kant does not reduce our emotive response to the beautiful and the sublime as something so purely subjective that it belies judgment. Neither does he leave us with the reduction of emotion or taste to the proposition that you have your sensibility and I have mine and there is nothing more to be said about it. Rather, he wants to offer us a critique of judgment that defends a specific form of judgment provoked by feeling but that is not simply overwhelmed by it.

Some interpreters of Kant have wanted to save him from the subjectivism of the palate. They end up reducing the *sensus communis aestheticus* to aesthetic conventions materialized in the common sense of an existing community such that one person's judgment of taste can appeal to these conventions to convince others that this or that object is beautiful or sublime.[30] Yet the principle ·of taste is a subjective principle because it refers to the feelings aroused in the individual through her experience of pleasure or pain upon confrontation with a beautiful or sublime object.[31] Admittedly, Kant does write: "We might even define taste as a faculty of estimating what makes our feeling in a given representation *universally communicable* without the mediation of the concept."[32] But a *communication* of aesthetic judgment is possible in Kant *not* because someone shares the existing aesthetic standards of a particular community—this would be a contingent and thus not universal judgment—but because we can *imagine* that others would join in if we all adopted an enlarged mentality.

The *sensus communis aestheticus* to which Kant refers always points toward an *ought to be* of a shared community: the enlarged mentality in which we might articulate to one another the subjective basis of our reflective judgment of the beautiful and the sublime, and find it illuminated through the viewpoint of the other and echoed in the other's attempt to communicate his or her feeling. The futurity of this community of the *ought to be* remains open as a possibility in the *sensus communis aestheticus*. The enlarged mentality to which Kant refers, in which we attempt to think from the standpoint of everyone else, does not turn us to everyone in a given community but to anyone who can be included in the idea of humanity. Even this enlarged mentality is not constitutive of reflective judgment but only suggestive of what is demanded by it. Still, Kant's definition of an enlarged mentality shows how far he is from appealing to conventionalism or a conventionalism corrected by elitism as the basis of reflective judgment. It is not a group of experts who represent a cannon of aesthetic wisdom to which Kant is appealing as the basis of an aesthetic judgment.[33] "But the question here is not one of the faculty of cognition," Kant writes, "but of the *mental habit* of making final use of it. This, however small the range and degree to which a man's natural endowments extends, still indicates a man of *enlarged mind:* if he detaches himself from the subjective personal conditions of his judgment, which cramp the minds of so many others, and reflects upon his judgment from a *universal standpoint* (which he can only determine by shifting his ground to the standpoint of others)."[34]

But what does Kant mean by "the standpoint of others"? He does not mean to be taken literally here, as if the task before us were to place ourselves in the shoes of an existing individual or actual group of individuals and see through their eyes. He does not mean for us to accept the standpoint of any given community—for example, all educated German men.

The others are imagined since they are ideal representatives of "the collective reason of mankind,"[35] which has no temporal closure. The imagination can help us in conjuring what might be the standpoint of the collective reason of mankind. In the end, however, each moment of imaginative representation is just that: a representation of what this *might* mean. In other words, there is no ultimate representation of the collective reason of mankind such that the imagination could bring the process of reflective judgment definitively to a close by basing itself on any of its representations.[36] The key here is to grasp that, by common sense, Kant means *feeling*, not something like the widely shared standards of judgment, or rationally deduced concepts under which we apply particulars. How can feelings exist in common? In a literal sense, they cannot, which is why the critique of judgment seems to involve antinomies. Feelings arise in the subject, and yet, if there is no possibility of communion with others, then there is only the subjectivity of the palate. So the question arises: what kind of judgment is consistent with the subjectivity of feeling?[37] Aesthetic judgments rely on metaphoric transfer, or what Kant calls *hypotyposis*, for the representation of communion. This communion must be presumed in the judgment. And yet, ultimately Kant concludes that this presumption of possibility cannot be rationally demonstrated.

When we judge an object as beautiful or sublime, we include the *should be* of the universal—that is inseparable from an idealized humanity—within the judgment at the time we make it. We do this without empirically consulting any other standards than our own, and without discussing it with anyone else before making the judgment. Thus, the *sensus communis* demands a particular kind of public sense that has nothing to do with what we normally think of as "community"—the set of standards composing an actual public:

> The name *sensus communis* is to be understood [as] the idea of a *public* sense,
> i.e., a critical faculty which in its reflective act takes account (*a priori*) of the

mode of representation of every one else, in order, *as it were*, to weigh its judgment with the collective reason of mankind, and thereby avoid the illusion arising from subjective and personal conditions which could readily be taken for objective, an illusion that would exert a prejudicial influence upon its judgment. This is accomplished by weighing the judgment, not so much with actual, as rather with the merely possible, judgments of others, and by putting ourselves in the position of every one else, as the result of a mere abstraction from the limitations which contingently affect our own estimate. This, in turn, is effected by so far as possible letting go the element of matter, i.e., sensation, in our general state of representative activity, and confining attention to the formal peculiarities of our representation or general state of representative activity.[38]

This idea of the public always implies an experiment of the imagination because we are called to imagine all the possible, not just the real, viewpoints of others. Thus, the *sensus communis aestheticus* always implies a public that awaits us, not one that is actually given to us, nor one that can be given to us, once and for all, in any predetermined public forum. The judgment creates the community, not vice versa: "The judgement of taste exacts agreement from every one; and a person who describes something as beautiful insists that every one *ought* to give the object in question his approval and follow suit in describing it as beautiful. The *ought* in aesthetic judgements, therefore, despite an accordance with all the requisite data for passing judgement, is still only pronounced conditionally."[39]

Since in reflective judgment we can never take a particular and simply subsume it under any given concept—that too would undermine the subjective basis for such judgment—we will often rely only on aesthetic ideas to communicate our judgment once it is made to others:

[B]y an aesthetic idea I mean the representation of the imagination which induces much thought, yet without the possibility of any definite thought whatever, i.e., *concept*, being adequate to it, and which language, consequently, can never get quite on level terms with or render completely intelligible.—It is easily seen, that an aesthetic idea is the counterpart (pendant) of a *rational idea* which, conversely, is a concept, to which no *intuition* (representation of the imagination) can be adequate.[40]

Put most simply, because we presuppose *a priori* a *should be* of universal agreement at the very moment we make a judgment, our judgments are not purely subjective. Yet Kant concludes that a necessary presupposition can never be directly proven by reason, only represented as an aesthetic idea. In reflective judgments, we thus cannot reach rational or even imaginative closure. This does not mean that we have no basis to judge; rather, it simply means that the mode of reflective judgment is itself aesthetic and demands from us an enlarged mentality, which can always be expanded further.

The sublime is evoked by the often heart-rending awe we feel, for example, before the great force of nature:

> In the immeasurableness of nature and incompetence of our faculty for adopting a standard proportionate to the aesthetic estimation of the magnitude of its *realm*, we found our own limitation. But with this we also found in our rational faculty another non-sensuous standard, one which has that infinity itself under it as unit, and in comparison with which everything in nature is small, and so found in our minds a pre-eminence over nature even in its immeasurability. . . . In this way external nature is not estimated in our aesthetic judgement as sublime so far as exciting fear, but rather because it challenges our power (one not of nature) to regard as small those things of which we are to be solicitous (worldly goods, health, and life), and hence to regard its might (to which in these matters we are no doubt subject) as exercising over us and our personality no such rude dominion that we should bow down before it, once the question becomes one of our highest principles and of our asserting or forsaking them. Therefore nature is here called sublime merely because it raises the imagination to a presentation of those cases in which the mind can make itself sensible of the appropriate sublimity of the sphere of its own being, even above nature.[41]

A judgment of the sublime is complicated for Kant because, to make such a judgment, we need to confront the force of nature with an attitude resembling the moral—moral in the sense that it is through an analogy with our freedom that we sense the sublime. This feeling simultaneously combines how small we are, in the natural scheme of things, with the inscrutability of our being irreducible to physical and contingent circumstances that induce us to judge an object, an encounter, or a person as sublime. Sublimity in this sense calls forth in us a feeling of respect for who we are as free beings at the same time that it overwhelms us as sensuous creatures.

Death of course is an obvious example of such an overwhelming natural force none of us can escape. My mother's courage in the face of death made it seem a small force indeed. But to challenge the power of death to render us helpless boggles both the mind and the imagination in exactly Kant's sense of the sublime. We know people die. That we can understand. But we cannot understand death, nor can we even imagine what death is. I have no adequate words for my mother's bravery as she exercised her right to die, but I can at least write that I was awestruck by her dignity. Yet, my mother, like Spivak's distant relative, depended on a witness to be faithful to the "truth" of her own death—the "truth" of her claiming her own person as the actual means of her death. My duty was to evoke the sublimity of her death for the reader. I was left with the task of representing the significance of what, on the level of imagination, is unrepresentable: death itself. I was called to this precarious task in my mother's last days. This entire book is my response to her naming me as her witness to her moral freedom. This book, then, also demands that the reader join me in the *sensus communis aestheticus* through the process of making an aesthetic judgment of my mother's exercise of her right to die as sublime.

As a dramatist, Schiller helps us understand why we need others if we are to bear witness to our moral freedom. We cannot do it alone. By evoking the *sensus communis aestheticus*, we are called to witness on behalf of the person or dramatic persona. Schiller places his own analysis within the Kantian framework, particularly in his emphasis on the aesthetic significance of moral freedom. But he begins by correcting Kant slightly concerning the issue of fear:[42] "Someone who overcomes what is fearful is magnificent (*gross*). Someone who, even while succumbing to the fearful, does not fear it, is sublime (*erhaben*). . . . An individual can display magnificence in *good fortune*, sublimity only in *misfortune*."[43]

Schiller goes on to make a further important distinction that Kant does not make. The sublime, Schiller suggests, always represents the presentation of three images: some objective physical power, our fragility as natural creatures before this power, and our moral strength, which leads us to confront it, our fear notwithstanding. These images can be presented in different orders but they all must be represented in their sublimity. But there is a distinction between two kinds of representations of the sublime. First, there is the *contemplative sublime*, which involves the representation of power and force only. We are left to imagine the suffering that would be endured by anyone confronting it, and how that confrontation would demand anyone's moral freedom. In the *practical sublime*, on the other hand, the force, suffering, and actual moral struggle of the person is represented. In the second case, the task of our imagination is to evoke an image of ourselves as suffering physical beings who might be called to exercise our moral freedom in exactly the way it is represented to us. When the image of moral freedom arouses our own emotion, suffering then becomes *pathetically sublime*. By *pathetic* Schiller does not mean pitiful. He instead means that the person on stage arouses in us the *dramatic pathos* inherent in that person's circumstance. Thus, by the *pathetically sublime*, Schiller means the dramatic pathos that implies moral courage or struggle in the presence of what is represented as awesome. *Dramatic pathos* makes us suffer with the person represented—either on stage or in some other manner—through any series of empathizing emotions (fear, anxiety, and so on). If it works, it is because of the feeling of awe or respect that the person's dignity, along with her suffering, evokes in us. In the *pathetically sublime*, we need both to imagine ourselves undergoing the person's suffering and to feel the emotions we project onto the representation. At the same time, we must allow ourselves to feel the strength of soul of the person as she claims her own freedom. For Schiller, it is not morality per se that is important in the *pathetically sublime*. Indeed, he believes it is dangerous to look for moral legitimacy there.[44] We sense the imagined moral struggle and moral resistance as they allow us to represent our freedom through an imaginary identification with what is being represented. To experience our freedom, we need to imagine the fight as if we, too, could take it on. We need to imagine the struggle. In this way, while the other person refuses to surrender her dignity and arouses in us the identification that frees our own imagination for a moment, we experience how we too *could* be free.

SPIVAK'S HISTORY AND THE PRACTICAL SUBLIME

Though Spivak does not use these terms, I believe her critical engagement with feminist history is inseparable from her dramatic presentation of her journey to find out what she could about the Rani of Sirmur and her effort to testify to what she saw as the ethical meaning of Bhubaneswari Bhaduri's politically motivated suicide.[45] My interpretation of Spivak is that the "miming"—her word—of her journey is a dramatization of the sublime aspect of what remains of these women, a sublimity that shows their own struggle to preserve something of how they identified themselves. Both women are buried under the legacy of imperialism. And yet, something of them remains. Spivak gathers the remains and gently blows the precarious ashes into their ghostly shape. She represents herself as she watches them take form in time. This representation of Spivak's own "moral love"[46] for her ghosts demands the imagined possibility of ethical resistance. Without ethical resistance, we would not be inspired to make the reflective judgment that the ghosts being represented to us are sublime. This imagined possibility can help us disclose a certain kind of feminist community—a community that can witness to Spivak's Rani, to Bhubaneswari Bhaduri reimagined at the moment of her death, and to Spivak herself as she stages this struggle in order to give these women dignity, since they "appear" in colonial history only as unworthy of respect. My support for this reading of Spivak comes in part from the initial question she asks herself when she frames her archival project: "If I ask myself, How is it possible to want to die by fire to mourn a husband ritually? I am asking the question of the (gendered) subaltern woman as subject, not as my friend Jonathan Culler somewhat tendentiously suggests, trying 'to produce difference by differing' or to 'appeal . . . to a sexual identity defined as essential and privileg[ing] experiences associated with that identity.'"[47] By asking the questions as she does, Spivak returns *pathos* to the ashes she gently handles. But despite all her care,[48] she cannot give us a fully coherent story. Too much has been lost of these women as they once were in their own environment. In the face of the extraordinary difficulty of her task, Spivak courageously takes responsibility for her representation.

By evoking exactly what Friedrich Schiller calls the *pathetically sublime*, Spivak relies on a dramatic representation of the cuts and the silences. "The image of another's suffering," writes Schiller, "combined with emotion and the consciousness of the moral freedom within us, is *pathetically sublime*."[49]

THE COMMUNITY OF THE OUGHT TO BE
AND FEMINIST WITNESSING

Spivak can represent the "fragile ghost" of the Rani of Sirmur only by miming her own journey in search of the actual truth behind the Rani. She calls on us to imagine her wandering around in what used to be the men's quarters of the Rani's palace, locked in as the Rani herself once was. This explains her unique writing style in her chapter on history. There is not enough of the Rani's character to allow Spivak to represent her more directly. So instead, she represents the cuts in her own journey, the points that allow us to see the Rani's resistance to the British, and then the abrupt end to her story when she ceased to be needed for their settlement. We imagine the Rani through Spivak's staging of her own journey in search of her, a journey that she places within the frame of a possible retelling of the meaning of *Sati* as an ideological battleground between feudal patriarchy and capitalism. Spivak's restaging of the ideological battlefield allows us to grapple with the notion that "such a death can be understood by the female subject as an *exceptional* signifier of her own desire, exceeding the general rule of a widow's conduct."[50]

If we cannot imagine that there was moral resistance, even of acts that took place within the context of the patriarchal formation of who the Rani could be, then we cannot feel the sublimity of her precarious presence. If we are to feel such fragility as a sublime object and, in doing so, commemorate her in the only way that is left for us, Spivak's speculative retelling of the scene in which *Sati* took place is crucial. Schiller reminds us that "poetic truth consists, not in the fact that something actually happened, but rather in the fact that it could happen, thus, in the internal possibility of the matter."[51]

If we can imagine with her the ethical significance of the person who might have been lost, then Spivak's search for the Rani of Sirmur brings poetic justice to the Rani's "fragile ghost." In this way, it is less difficult for Spivak to return sublimity to the ethical suicide of Bhubaneswari. As Spivak makes clear, Bhubaneswari tried to "speak" with her body. But even her family could not read her action other than in a feminine way and context, which denied its sublimity. Spivak uncovers a story that classically fits Schiller's definition of a sublime object: a physical force that Bhubaneswari feels compelled to resist—colonial oppression—suffering before her inability to carry out a task she believes is politically necessary, a task that is avoided in a nevertheless ethical way. Only if we "join" Spivak in a *sensus communis aestheticus* and make the aesthetic judgment about the sublimity of the "stories" she tells us, will the feminist witnessing to which Spivak calls us be brought into being.

When confronted with an object judged as sublime, we are turned back to our awe and respect for our freedom. To communicate our feeling of the sublime, then, we must always evoke our freedom. This is why without Spivak's speculative restaging of the scene of *Sati* we cannot possibly feel the sublimity of the fragile ghost of the Rani of Sirmur. Still, the *sensus communis aestheticus* cannot be anchored either in the world of nature or the world of freedom. If it is to remain faithful to Kant's own conception of reflective judgment, then it cannot be *anchored* at all. By making such an aesthetic judgment, the feminist *sensus communis aestheticus* arises as we bear witness to Spivak's history. Our community is that fragile. When Spivak appeals to us to imagine the Rani of Sirmur locked in her palace, struggling to resist the British intrusion into her life, she also asks us to join her in this community.

If we can imagine neither the sublimity of the ghost of the Rani of Sirmur, for whom Spivak takes responsibility, nor Spivak's tribute to the dignity in Bhubaneswari's suicide, then that community crumbles upon the failure of our moral imagination.

Within Bhubaneswari's own family, it was this very community that tragically failed to arise: "Bhubaneswari had fought for national liberation. Her great-grandniece works for the New Empire. This too is a historical silencing of the subaltern. When the news of this young woman's promotion was broadcast in the family amidst general jubilation I could not help remarking to the eldest surviving female member: 'Bhubaneswari'—her nickname had been Talu—'hanged herself in vain,' but not too loudly. Is it any wonder that this young woman is a staunch multiculturalist, believes in natural childbirth, and wears only cotton?"[52]

THE RELEVANCE OF SPIVAK'S HISTORY FOR CONTEMPORARY FEMINISM

A feminism that fails to come to terms with our mother's stories loses its soul and disorients us. We fall into the now of advanced capitalism, the quick cure. That Bhubaneswari's suicide was part of her struggle for national liberation is also part of Spivak's own family history and indeed inspires her feminism. The failure to heed its significance is part of the feminism of her ancestresses who now work for the "New Empire." Who we are can never begin just with the present. In any political position we take, we are constantly working through our own identifications; we are assuming responsibility—or refusing it—for the phantoms that inevitably haunt a patriarchal history that has denied so many women their significance, their voice. The young "emancipated girl" does not know what she is up against, nor how easily her "emancipation" can be co-opted. Spivak calls us toward a "women's time" to remind us not only that what we forget we are fated to repeat, but that we are inseparable from our own reclamation of those obscured by "official history," and that another history can be spun into another story only by respecting another time. In her words, "historically, legitimacy was of course established by virtue of abstract institutional power. Who in nineteenth-century India could have waited for women's time here?"[53]

Women's time is not just about the actual, arduous journey Spivak undertook to find the traces of the Rani of Sirmur. For her journey can be understood as an allegory for how difficult it is not only for historical voices that have been suppressed but also for new voices to find the means of representation to be seen and heard. Women's time is the generational time in which the groundbreaking work of organizing can take place and consolidate the changes it brings about. The tendency to rush in for the quick fix—the one that surpasses the organizing efforts going on throughout our world at the grass-roots level—will once again fail to heed the gendered subaltern. We are called to bear witness to women's suffering under injustice and to the silence effectively imposed upon women's stories as a matter of history. Given the horrifying realities of the injustice imposed upon women and men throughout the world, we are called to do this all the time. Often designated under the code word *globalization*, the "New Empire," as Spivak calls it today, is a nightmare for many of the world's people. There is much resistance to the horrible inequalities it has brought in its wake. Often these struggles are represented in all their sublimity to those of us residing in the West. This is true of the historic struggle of the El Salvadorian National Liberation Front to continue their fight against the military might of the United States, the long endurance and resistance of the Palestinian people, the extraordinary victory in South Africa to overthrow apartheid and constitute a new nation-state. Yet it is not just the brave soldier who is sublime. Students who run into the line of fire to protest torture have represented themselves in all their courage. Nor does action need to be so dramatic. A group of workers who stand up for themselves in a strike and refuse to back down evince their sublimity in the way they represent themselves. When I was a union organizer, one of my own most powerful experiences was my participation in a wildcat strike. All of these struggles take place within their own terms and within their own conceptions of what is at stake.

My point here is that Westerners can easily miss the sublimity of these struggles and thus become completely confused about what kind of support is being asked of them—for example, in a transnational labor struggle. Implicated as it is in our imperialist history, our social imaginary can easily block our sense of the sublimity of these struggles as they are being represented to us. When we unknowingly do not accord the participants in those struggles the very dignity and respect we presume for ourselves, the fragile *sensus communis aestheticus* in which we might feel totally overwhelmed by the representations of those struggles cannot form. If our moral imagination fails us in our effort to bear witness, then we undoubtedly will go astray in the manner of "aid" we choose to give.

Dignity is a paradoxical word. When we pay heed to the dignity of others, we do it for the sake of their immeasurable singularity. And yet, the unique value we impute to them makes it seem as though they alone carried the infinite worth of all of humanity. In this way, Hannah Arendt is correct to amplify the contradiction at the heart of Kant's notion of dignity: "In Kant himself there is a contradiction: Infinite Progress is the law of the human species; at the same time, man's dignity demands that he be seen (every single one of us) in his particularity and, as such, be seen—but without any comparison and independent of time—as reflecting mankind in general. In other words, the very idea of progress—if it is more than a change in circumstances and an improvement of the world—contradicts Kant's notion of man's dignity. It is against human dignity to believe in progress."[54]

What does Arendt mean when she argues that progress is against dignity? She is being faithful to Kant's articulation that the basis for our dignity resides in our freedom. As a matter of principle, we can all act freely. With the exception of the specific sense Arendt elaborates, an argument for progress resolutely denies that we—and here *we* refers to all who can be imagined within the inclusive reach of humanity—are, *in principle*, free. In Kant, freedom is inscrutable because if it could be reduced to a set of positive characteristics, it would no longer be freedom but would come to be known as a cognizable object falling under the laws of nature. Beyond the inscrutability of our freedom, dignity and the respect it demands do not require the elaboration of common identities and shared characteristics. If we put human beings on a scale through an elaboration of these characteristics and try to measure progress by regarding one group as more "civilized" than another, we ironically deny the dignity of the very people we are "weighing," even when we are giving them a positive "grade" on the progress scale. The ethical meaning of Arendt's insight into the contradictory relationship between progress and dignity is simply this: when we seek to render our world more just, we must do so in accordance with the respect for the dignity of all others. In our own political struggle for justice, the ethical lesson concerns how we must treat others. That someone is not "high" enough on some value scale that attempts to codify the meaning of progress[55] is never a justification for wronging the dignity of another person.

The respect that dignity demands actually serves as a moral barrier to the translation or the appropriation of the other into our various social and cultural frameworks. It also provides Spivak's history with the ethical force of sublimity, to which we are called to witness.[56] To bear witness to sublimity we have to accept the infinite worth of the other expressed in her dignity. It is one thing to recognize that we cannot know the other's experience and thus validate it in cognitive terms, and another to witness to the gendered subaltern. Often we are called to admit that we cannot know the other in her difference. But the recognition of difference all too easily degenerates into indifference. Instead, Spivak asks us to bear witness to the sublimity of the subaltern and thus to their infinite worth. In this judgment, we are called to accord them the infinite worth that they have always been denied in the so-called "civilizing mission" that devalues them as the "natives." As Western feminists seeking to find what role we have to play in transnational reform efforts, dignity demands of us that we attribute not value but infinite worth to our sisters in struggle. It not only slows us down; it stops any attempt to value or devalue them as victims in need of our help. Perhaps at this slowed-down pace[57] we can witness to the sublimity of what has almost never been seen as sublime— the day-to-day endurance of women who sustain, against all odds, their struggle to change the world. If we have the courage to stay in women's time and to place ourselves under the mandate of respect, we may be able to dream up new possibilities for solidarity not ensnared by our imperialist legacy.

Chapter IV

COOPERATION FOR DIGNIFIED LABOR:
MOVING TOWARD *UNITY*

> *When a girl comes to this world and a doctor takes her out of this, you know, the womb of the mom, and says, "This is a girl"* . . . *I remember when I had my girl and the doctor told me, "It's a girl," I cried. And I didn't cry during labor. Even with all that pain, I did not cry. But I cried when he told me "It's a girl." There is a name for it: a hero. It's a hero. It's a hero! Because when a woman comes to this world she will have to live in this world, and do so many things that a man will not do. Women will. Nobody else. So that was a hero* . . . *who came out of me. The first hero I met was my mother.*

—Zoila Rodriguez, *Administrative Assistant,*
The Workplace Project

*J*ust as the adoption of my daughter Sarita was a sublime experience, so was the birth of Zoila Rodriguez's "hero." Day-to-day moments in life can become sublime and can lift us together into a community of the *ought to be.* The sublimity of others' struggles demands an encounter with their dignity that dismantles our own guards against identification with them. Sublimity can manifest itself as an epiphany that takes us out of ourselves and makes us see what previously remained invisible to us.

Even when we are not transported by the identifications opened to us by a sublime experience, we can, however, build "unlikely coalitions"[1] and fight the crucial battles against injustice. In this chapter, I return to one of my main themes: the central importance of dignity as an ethical demand placed on Western Anglo feminism. I will speak of my own journey to develop a relationship with *Unity*, a cooperative of housecleaners in Long Island, New York. In the next chapter, some of the members of *Unity* will tell their own stories about how they came to this country from El Salvador, and what their work life was like prior to their involvement with *Unity*. Dignity is built into the organization of the cooperative. The struggle for self-representation and self-organization is a difficult practice involving communication, education, and political action. This practice can get off the ground only if the participants meet each other as equals, each entitled to this treatment as a matter of her dignity. As Zoila Rodriguez emphasizes: "We are asking for what we deserve already. That's what we are claiming. We deserve this. We deserve a better life: a life with dignity . . . because every human being is born with it."[2]

Each member is respected in her capacity to represent herself and the needs and interests and dreams she believes she shares with others as the organization slowly comes together. She is viewed by others as assuming responsibility for herself and for her relationship to the group. Responsibility is emphasized in workshops, which every potential member must complete as a condition for joining the cooperative. Self-representation, and with it, the assumption of responsibility for the person one is becoming, is not simply a matter of self-image for the members of *Unity*. It involves them in acts of interpretation and communication with others as they try to define and redefine a sense of themselves and their basic identifications.[3] *Unity* teaches us that dignity demands respect for the self-representations of others. Sometimes, in recent political theory, the elaboration of ethical principles is seen as a limit on real participatory democracy.[4] But in *Unity* and *The Workplace Project*—the umbrella organization that facilitated the development of *Unity*—dignity is a principle of self-organization that demands truly democratic participation. *Unity* requires that all employers who hire members of the cooperative respect their work as dignified labor, to use Martin Luther King's famous expression. "For example, one of our fellow workers was told to clean the kitchen floor, which was very large," says Zonia Villanueva in her interview, "and they wanted to her to get on her knees and scrub it with a brush, not a mop or anything. . . . So she did this the first time, and then she asked the lady for a mop. Well, she didn't give her the mop or anything, so she told her that she couldn't continue working like that. So we called and cancelled because what we want is for women not to get on their knees. . . ."[5]

Unity reinforced my understanding that respect for the dignity of all women is absolutely necessary if we are even to begin the arduous task of a multicultural or intercultural exchange. White Anglo feminists are being challenged because of their failure to see how claims of "global sisterhood" can replicate the divide between the so-called civilized, free West, and "all the rest in the world" who supposedly need us.[6] Sometimes, in place of listening, we impose what we *imagine* as necessities for women's liberation, as if our felt necessities were those of everyone else. Even with the best intentions, we often cast ourselves as the new rescuers who must go "over there" to help and rescue our hapless sisters in the Southern nations.[7] The ethical vigilance inherent in respect demands that we interrogate the Orientalist and imperialist frames of reference that often render us blind and deaf to the visions and voices of women of color throughout the world. A crucial aspect of "Americanism" is that the United States is the "greatest nation in the world." Again and again, the United States government has used the pretense of concern for human rights as a justification for military action against other countries.[8] Feminists of the United States must be particularly on guard against their own investment in Americanism. We are often put in positions of power in NGOs—Non-Governmental Organizations—and other human rights organizations simply because we are citizens of the United States and/or associated with institutions explicitly or implicitly dominated by U.S. foreign policy.

Respect for the dignity of others does not mean that we must forego critical engagement *with* them. Refusing that kind of engagement with others is simply condescending. Respect for dignity requires us to reflect about how we are positioned in the dialogue. Without this reflection, we will be unable to take on the struggle "to become fluent in the histories of others."[9] Respect insists on the equality of all peoples as we try to build alliances and raise consciousness about our habits of thought and dispositions toward the world that we do not question. All of us come into this world as always already represented. We not only come out of different cultures, classes, ethnicities, nationalities, and languages, but there are clear privileges that some of us have because of those identifications. Regrettably, sometimes that privilege is simply erased under the pretense that white Anglo women are in a special position due to their greater knowledge of what women need and can achieve. We forget that this representation of ourselves is part of the "troubled legacy" of imperialism.[10] Gayatri Spivak constantly reminds us that knowledge grows out of contact and indeed conflict with difference.[11] But knowledge, including greater apprehension of what we do not know, can grow only out of a process of questioning our self-representations and recognizing how they are enforced as legitimate, betrayed by stereotype, or altogether pushed under representability by the dominant culture and society. Put most simply, critique among equals presumes respect. The dignity of others demands that white Anglo feminists get off their high horse and stop representing themselves as the civilized rescuers of women in other parts of the world desperately in need of help. We must dare to "get down" and have our own representations of our reality and ourselves challenged. Otherwise, we will be left looking down for someone who was never there in the first place, but who came into our view only as we imagined them from a position of class, ethnic, and racial privilege.

IDENTIFICATION, POSITION, AND REPRESENTATION

Let me now introduce a distinction I believe can help us understand the complexity of our actual lived identities. That distinction is between position and identification. I am positioned as a white Anglo woman in this society. I am positioned from my passport and other legal documents as a citizen of the United States, and I am certain that this position often allows me to cross *some* borders without a second glance on the part of immigration officials.[12] When my adopted daughter Sarita first traveled with me to England at the age of three, she was a Paraguayan citizen only. Paraguay was on England's "suspect" countries list. She was initially asked to step out of line to be interrogated. After it was noticed that she was only three years old, a long discussion ensued between officials of British Airways and U.S. Immigration. We were allowed on the plane. We were also given a piece of paper we hoped would get us into England without Sarita being stopped and interrogated. I tell this story because Sarita's passport positioned her in terms of her relationship to both governments. My passport, on the other hand, *positions* me as a citizen of the United States before immigration officials when I travel regardless of how I might *identify*. In my heart of hearts, I identify as Scots-Irish.

My identification as Scots-Irish has never let me rest easily with the position of white and Anglo. After all, for me such identification has a meaning strongly linked to the brutal history of English imperialism against the struggle of the Irish for independence. For some of us who identify as Irish, our identification carries with it an explicit rejection of the idea that we are English or "Anglo" in any way. That English is the language of Ireland and that someone who is Irish is an English speaker resulted from the appalling suppression of Gaelic, the Irish language.[13] Many people struggled to the death for their language. But in another context, another scene, place me before immigration officers, passport in hand, and no matter how I choose to identify, I will be positioned as a U.S. citizen and as white and Anglo.

Identifications are fluid because we internalize them and, consciously or unconsciously, reinterpret their meanings as we make them our own. This is why we are in the end responsible for our identifications. Even if they clearly come with sedimented meanings, they are never simply imposed on us. A position is harder to change because it immediately involves us in the social and symbolic networks that make up our social, cultural, and material world. Of course, the two—position and identification—are related, and an explicit identification can be a demand for both collective and individual agency. But positions cannot always be dissolved through acts of identification.

To change our positions and the privileges or oppressions that come along with them often demands political, ethical, and material transformation beyond what any one of us can do alone. Yet, even by ourselves, we can still take small steps in that direction. We can acknowledge our own identifications since these allow us to assume privileges for ourselves on the basis of race, class, and ethnic position. But the effort to acknowledge our identifications and to change the meaning they have for us does not mean we can simply step out of our positions. I am white in this society no matter how I identify myself.

For me, an ethical identification calls for the recognition that we are in a privileged position. I am positioned as white. Paradoxically, whiteness masquerades as an erasure that is in fact a color. If one looks back into the history of the United States, "colored people" were always the ones with the "color" in our society. To identify as white is to make the point that whiteness is an identification. The meaning of whiteness is inseparable from the meaning it has been given through the history of white supremacy. Even if I do everything I can through anti-racist politics to reject the racism passed down in the word and the identification, I will still be seen by others as white as I walk down the street.

The ability to speak English has been turned into an identification: English speaker. This identification often defines itself through the effort to suppress Spanish.[14] To identify ethically as white and Anglo in the United States in the twenty-first century is to recognize this openly as a position that continues to carry privilege. We are never reducible to our positions. Nor can we simply separate ourselves from them. We can begin to question the meaning of our position only if we first acknowledge it as our own. This questioning is the kind of ethical vigilance to which true respect for the dignity of all others calls us, and which allows us to begin the struggle to name and understand "translatable universals."[15] The practice of *Unity* shows how this cooperative has translated dignity as a universal ideal into the organization of the cooperative. However, I learned about the translation of this ideal only in the course of our conversations. So we need to take a step back. Let me turn now to my own engagement with *Unity* and to our shared effort to achieve a multicultural and intercultural dialogue.

WHY *UNITY*?

I first went to *Unity* to interview a "nannies" cooperative. The second wave of feminism, at least during its early days, placed the demand for decent subsidized childcare at the center of its agenda. Socialist feminism had always struggled for the public sponsorship of what had become privatized labor in the home and what was thus designated as women's work.[16] The argument in the end was simple: women's private labor in reproducing and raising children should be publicly supported so that women can enter and participate in all aspects of public life. The hope was that once it was publicly supported, it would lose its gender basis. Study after study has shown that, despite social services, the actual burden for raising children remained on women's shoulders in those societies that called themselves socialist.[17] Still, it would be foolish to think that the provision of these services did nothing to help women enter the work force.[18] The lesson instead should be that the provision of social services is not enough to disrupt stereotypical gender roles. The loss of those services has had a serious effect on women's lives in what used to be the "actually existing" socialist countries of Eastern Europe in the Soviet bloc. Privatization has not done much good for women.

In our own country, we were starting from scratch when we first demanded childcare. The history of publicly funded childcare is sketchy at best; centers were often established only as an emergency measure in order to bring desperately needed women into the labor force, as was the case during World War II. The kinds of programs that should have been developed had to be debated, along with the need for such programs in the first place. There were fierce debates over whether employers or the state should fund these centers.[19] If these centers were subsidized, there were battles over who would actually control the decision-making process in the centers. In minority communities, the issue of community control was at the heart of the matter. Truly democratic processes for community control were fought for, and in some communities they were established through boards governed by parents and teachers. The battle over who would care for the children and how they would be cared for was understood to be key to the struggle against forced assimilation.[20] Children were seen as the future. It is now often forgotten that the Black Panther Party became well known in Oakland largely because of its breakfast and lunch programs. Throughout the 1980s, the battle over community control and the effort to maintain centers with ever-waning resources went on against greater and greater odds in minority communities.

But the fight for subsidized childcare seemed to lose its steam in the "women's movement." There is no doubt that the loss of energy was a result of the full force of the opposition during the Reagan years. The Republican Party developed its strike-back agenda against feminism in a series of slogans,[21] one of which was "family values." Feminists who continued to call for childcare were accused of being selfish, narcissistic women who did not want to take up their proper role in life. Childcare was attacked as unnecessary and dangerous to the well-being of children. Children needed mothers and not institutions, or so the argument against childcare went. Yet these accusations were completely out of touch with reality. Work is not a luxury. For most women in this country, it is an economic necessity.[22] In the 1970s and '80s, the successful legal challenges against women's employment in the traditionally masculine jobs opened doors that had previously been slammed in women's faces.[23] Women began entering the professions from which they had been barred. In spite of the backlash that feminists have had to endure, we sometimes forget how much has changed in the last thirty-five years. There are now two women justices on the Supreme Court. Neither one of them could even hope to get a job as a lawyer when she graduated from law school, even though both made it to the top of their class. Still, it was not just new opportunities that were bringing women into the work force. More and more women who were mothers of children under five were entering the work force because there was no other choice. Some of these women were single mothers. But many others were married and entered the work force because two incomes were necessary to support their families. And yet there was no decent childcare to be had. The Republican Party's talk about family values was empty.[24] It had no intention of providing the support necessary for families to sustain themselves. Unlike the Republicans, Hillary Clinton tried to make childcare and education generally a rallying call for the Democrats.

In her formulation, it takes a village to raise a child.[25] But her call—like Clinton's healthcare programs—was blocked by the Republican Congress. The question concerning how women who work take care of their children is, then, a natural one to ask.

We have only to walk down any street in New York City to begin to get an answer, at least for some women, particularly those white Anglo women who have entered the professions in the big cities. As one walks up and down the streets of New York, one sees innumerable women of color walking or holding white babies or playing with older white children. Who are these women? Why did they come to the United States from other countries? How do they live? Where are their families? What are their stories? These also seem natural questions to ask in the name of the sisterhood that is supposedly the hallmark of feminism. And yet, these women and their stories have not been heard.[26] Truth be told, an overwhelming number of professional women are able to work only because the women of color who take care of their children are like "second mothers" to them.[27] It is one thing to talk about the harmful effects of globalization on women "over there" and to fight for human rights programs to better their conditions. Of course, that work is important,[28] but the reality is that those same forces of globalization provide professional as well as middle-class white women a cheap and easily exploitable work force.

The disastrous effects of U.S. foreign and economic policy can be seen and felt in our own home cities. Here I will only summarize arguments that have been eloquently and tirelessly made by innumerable economists and political scientists. Particularly as they support a specific kind of development in the nations of the global South, U.S. policies help create a surplus labor force. The development of a primarily export economy, which is maintained by brutally exploitative sweatshops, undermines the internal consumer base to support further growth. Workers are much too poor to buy the goods they produce. The exports are purchased at such low rates that they cannot begin to help the country pay off its debt to the International Monetary Fund and the World Bank. The emphasis on export-focused "free trade" zones in loan agreements often overrides all the "home countries'" labor laws, preventing the creation of a solid working class with rights and benefits.[29]

The huge debts themselves lead to significant downsizing, which in turn undermines the creation of a middle class and the development of an internal consumer base for the purchase of products produced within the country itself. Social services are curtailed or privatized in the desperate struggle to pay off the debt. Imposed privatization then leads to further layoffs. The infrastructure of the rural economy breaks down because it is not supported by the loans. Men and women are forced to go to the cities to find employment. This endless scramble, which goes around in circles, provides the surplus labor force.

Women workers continue to enter the economy in larger numbers as family sustenance farms are increasingly replaced by agrobusiness, whose focus is also on production for export. Some jobs, particularly those in manufacturing, are specified for women only. Just think of some of the assembly jobs in the electronic industry both worldwide and in the United States. The typical argument in this regard goes something like this: women's nimble fingers and ability to endure deadly boredom make them alone suitable for this kind of work. In many of the world's countries, women make up the majority of sweatshop workers because of the way gender is used to define certain kinds of labor markets.

In our country, both men and women from the global South provide necessary labor in some of the most marginalized areas of the economy. Far from stealing jobs from "Americans," they take the unwanted jobs that American workers themselves do not want.[30] Labor market segmentation and the lack of legal status keep them from even having the hope of breaking out of the secondary labor markets.[31] At the same time, the gentrification of the middle classes, through the booming finance industry in the 1990s, created a new group of wealthy younger people whose lifestyles led to the return of "servants" in a form different from the classic suburban middle-class family. As Saskia Sassen explains:

> The expansion of the high-income workforce in conjunction with the emergence of new cultural forms has led to a process of high-income gentrification that rests in the last analysis on the availability of a vast supply of low-wage workers. This has reintroduced—to an extent not seen in a very long time—the notion of the "serving classes" in contemporary high-income houses. The immigrant woman serving the white middle-class professional woman has replaced the image of the black servant serving the white master.[32]

Here we see how precarious the idea of "global sisterhood" is. Many feminists who employ domestics look away from the human rights violations that are occuring under their own roof.[33]

The case of Zoe Baird brought these "invisible" yet absolutely crucial workers into the public eye. When Zoe Baird lost the nomination for attorney general for hiring two "illegal immigrants," she tried to defend herself: "I was forced into the dilemma to care for my son. . . . In my hope to find appropriate childcare for my son I gave too little emphasis to what was described to me as a technical violation of the law."[34]

Many feminists defended her: "Colored" illegal nannies were the only solution out there. Zoe Baird should not have been singled out when so many others did the same thing. In the absence of publicly funded childcare, what else was there to do? After the Baird debacle, some feminist organizations called for a reform of the law that turned good, middle-class professional women into criminals. Two different kinds of reform bills were proposed. The first worked in the following way: A household employer would identify a domestic position, swear by affidavit that there was no U.S. citizen or legal resident available to fill the position, and petition to hire an "illegal immigrant." The second required that the Department of Labor determine when there was a shortage of citizens or legal residents who were available for the position of domestic worker. If such a shortage were found, then it would be possible for a working domestic employee to file for a visa. Note that the applicable word here is *visa* and not an application for permanent residence. In this process, an applicant could qualify if she demonstrated that she had worked in the home care industry for a certain period of time. She then had to swear that she intended to stay in that kind of work for a certain length of time. The second approach had one advantage: it allowed the worker to apply for a visa. But the problem was that she had to project continual employment. Under the first program, the employer was directly in control. Under the second program, the employer was indirectly in control. Neither program was perfect, although there is still discussion about their merits.

In the Spring of 2001, the bill that would give Salvadorian refugees an expedited process for permanent residency was once again turned down. *Unity*'s solution is more direct. The demand of *Unity* and *The Workplace Project* is "Unconditional Amnesty Now!" for all illegal immigrants currently residing in the United States. They include all workers in their demand; not just Salvadorian refugees. After the issue of low-paid immigrants in domestic positions rose to public notice, it died down, until the case of Linda Chavez brought it into the public eye once more. Linda Chavez was nominated to be Secretary of Labor in George W. Bush's cabinet. It was discovered that Chavez had an "illegal immigrant" living with her. Chavez described her relationship to her as one of compassion. She said she did not have an official employment contract of any sort with the immigrant, but instead, out of the kindness of her heart, supplied her with an allowance.[35] Undoubtedly, this is an extremely difficult issue for feminists to face. But engage it we must. There is no more intimate evidence of the class, ethnic, and racial privileges that divide us. After all, it is almost always white women who are the employers and thus the perpetuators of inequality among women. This inequality is no longer "out there" but in our own homes. It is not a defense against injustice that everyone else is doing it, too. Immigration under the present conditions of globalization has necessarily become a feminist issue.

The Workplace Project was an organization specifically formed on the basis of the knowledge of how globalization created a surplus labor force that could not escape the secondary labor market.[36] Since its 1992 establishment in Hempstead, Long Island, the intent of the project was to focus on facilitating the self-organization of immigrant workers. The campaign that first brought *The Workplace Project* rightful recognition was its effort to raise the wages of mostly male day laborers. Over three years, workers' committees were set up to organize the workers in order to demand higher minimum pay. In three years, and through the vigilance of tireless day-to-day organizing, the minimum wage for day laborers was raised by 30 percent. Almost all these workers were illegal. The transient nature of the worker meant that the workers' committees had to be constantly re-formed as new workers came to the corners looking for work and senior workers left.

Prior to the establishment of *Unity*, *The Workplace Project* began the "Domestic Placement Agency campaign." Jennifer Gordon, founder of *The Workplace Project*, describes her struggle to organize domestics. As with all workers who work in individual homes, it was hard to find a way to bring them together. Due to the intimate setting, many domestic workers feel close to their employers in a way that masks the class and exploitative nature of the work. Unions have traditionally been reluctant to organize these kinds of workers. What it means to be a worker and to see oneself as a worker is part of the struggle. Indeed, for some workers, unions have played a negative role in actively excluding some of them from the organizable working class. In her study of jute workers in India, Leela Fernandes explains the role of unions in promoting what counted as a class interest:

> Even as unions provide an important resource to workers, they have created their own bases of power contingent on social inequalities of gender and community.

In effect, the representation of "working class" interests through trade unions is constituted by hierarchical social relations that exist among workers. An analysis of trade unions thus cannot be divorced from an understanding of the differentiation of interests within the working classes. In this process, the boundaries of class are both produced and contested by the politics of gender and community. The differences that persist within the working class are manifested not by a pluralistic set of class identities but by a political process in which class interests are articulated through conflict, hierarchy, and exclusion.[37]

Fernandes's insight is not limited to the role of unions among the jute workers in India. Unions in the Unites States have long been guilty of producing rigid class boundaries and defining class interest within a framework of discrimination against minorities and women.[38]

In the last few years, the new leadership of the American Federation of Labor and the Committee of Industrial Organization (AFL-CIO) has committed themselves to organizing the unorganized. Organizations such as *The Workplace Project* are not meant to be an alternative to unions. But the understanding that class issues have both a sociological and cultural dimension and that what constitutes a class issue is itself a matter of political struggle allows us to see the relations among class, gender, race, and ethnicity differently. Whether something is represented as a woman's issue or a class-based issue is not something just in the nature of things. In her careful study of the jute worker, Fernandes makes this point again and again. Her argument is that gender hierarchies affect how class issues are formulated. Unions in turn marginalize the struggles of women workers:

> [H]egemonic discourses are translated into everyday practices that then produce particular kinds of gendered hierarchies and exclusions. One high-level management representative elaborated on these factors at length and then finally asked me with a measure of annoyance, "If there is a large supply of male workers, why should we hire women?" Implicit in his response is the assumption that women form a reserve army of labor that can be relegated to the private sphere of the home, and that women's employment needs are subsidiary to men's—an assumption that discounts both the subsistence needs of women in households headed by females, and the importance of the economic contribution of working-class women to the survival of the working-class family. Thus the "labor-market" is constructed through a gender ideology embedded within cultural codes of power and hierarchy.[39]

Organizers of the domestic placement agency campaign were well aware that they needed to develop innovative forms of organizing that accounted for the gender hierarchies in existence among workers who had become members of *The Workplace Project*. Most obviously, there were still clear divisions between men's and women's work based on stereotypical notions of gender. Men in the Hempstead, Long Island area who are immigrants from South America are mainly day-workers who do heavy work—for example, as gardeners or temporary workers on construction sites. It is women who almost exclusively work as domestic workers. In the case of women immigrants, the growing "feminization" of many of the low service jobs has brought them into the labor force. It has also pushed them into such blatantly exploitative jobs that they have assumed leadership positions in workplace organizing efforts—positions they could not have otherwise imagined as appropriate for women.[40] M. E. A., one of the women who tells her story in the next chapter, became a leader in the factory where she worked before she came to the cooperative.

The Workplace Project formed a women's committee precisely to recognize the way in which gender stereotypes affected the perception that women had of themselves and their role as workers. It was formed to give women workers a chance to represent their own interests and needs and to make sure that these were integrated into *The Workplace Project*'s organizational efforts. Organization and representation are not separated in *The Workplace Project*.

It was this committee that assumed the task of proceeding with the campaign. The initial research showed that many domestic worker agencies were charging twice the legal fee and advertising jobs at less than minimum wage. A large number of agencies—almost all of them—did not provide any contract between the worker and the employer. Hours were long, responsibilities unspecified, and the worker could be fired at any time if she did not comply with her boss's latest request. So what was to be done? Here is Jennifer Gordon's response:

> Our women's committee got together and essentially laid out a platform of what they wanted the agencies to agree to do. First, that they would only charge the legal fee. Second, that they wouldn't place women in jobs that paid less than the minimum wage. Third, that they would require both the worker and the employer to sign a contract that set out the working conditions and wages. Everyone would sign, and the domestic worker would have a copy of the agreement. It would be the first time that there would actually be a handle on what domestic workers are supposed to be doing.

In July 1997, we launched that campaign. Working with the support of labor and community and religious allies that we have around the island, after about nine months, we were able to get five of the six agencies on the island that we targeted to sign the statement of principles by pressuring them. We did this by phone calls, letter campaigns, basically by pressuring the agencies through surrogate clients. We would take people in a church or people in a community group that supported us and have them write letters or make phone calls or send faxes saying "I won't hire from your agency until you adhere to the statement of principles." The church or organization would likewise say, "We will tell our members, four thousand, three hundred or fifty members, that they should not hire from you until . . . Long Island CAN, the Long Island Progressive Coalition, and all sorts of groups around Long Island helped us put that pressure on the agencies."[41]

The results of the campaign were mixed. One difficulty was that enforcement was becoming a full-time job. One solution to that problem was to turn to the law. The problem was handed over to the Department of Labor and the attorney general. The response of both the Department of Labor and the attorney general were disappointing. This was hardly surprising since the exploited workers are considered "illegal" and, as a result, have no traditional form of political clout. Many workers who are not legal residents are also reluctant to participate in the legal institutions of the United States government. The other idea was "if you cannot beat them join them, differently." Rather than try to organize a competitive agency, *The Workplace Project* set out to form a cooperative to show that housecleaning could be dignified labor. And so the history of *Unity* began.

PROFANE ILLUMINATIONS:
WHY I WAS BROUGHT TO *UNITY*

I came to *Unity* with an understandable feminist purpose. I wanted to hear the stories of the women who worked as nannies. I wished to hear from the other mothers, those who were "second mothers"[42] to other women's children. I wanted to hear the voices of the women who had to leave their own children behind. I was aware that I also had another purpose: I wanted to know what were fair working conditions for "nannies." While interviewing Zonia Villanueva, the first thing I found out was that the cooperative's position regarding childcare was that privatized childcare could not be fair. All the women I interviewed had been nannies, and all maintained that those were the nightmare days for them. Once in *Unity*, they were determined not ever to go back to those jobs under those conditions. *Unity* was set up as a co-operative for housecleaning, period. "I feel that it is a dignifying job," Zonia said to me, "I set my own schedule. It's three or four hours. It's sixty dollars and I go home, and I have time to pick up my daughter. I have time for whatever else I want to do at home."[43]

Participation in the cooperative is time-consuming. Since *Unity* is a true cooperative, running the organization involves all of the members in major decision-making processes. All members have to serve at least in one committee that makes the cooperative run. Thus, these women work as housecleaners and as active members of the cooperative. The whole point is that they control the conditions of their work. Nannies, on the other hand, are controlled by their work and indeed sometimes subjected to the whims of their boss.

So I had my answer: the position of nanny is unjust. One part of me knew that because, for some time before I met the women of the *Unity* cooperative, I had been struggling to move away from anything that might resemble the usual nanny arrangement. But another part of me was struck by the answer. I began to realize that a whole series of deeply personal issues had led me to the cooperative. The interviews created a "profane illumination" about why I was there. Walter Benjamin defines a "profane illumination" as "crossroads where ghostly signals flash from the traffic, and inconceivable analogies and connections between events are the order of the day."[44]

I felt signaled. I felt drawn back. I was drawn to a story my grandmother had told me years ago. To remind the reader, my great grandmother worked as a maid when she first came to this country. For a while she worked as a private domestic and nanny. Sometimes Nana had to go help her mother after school. Once, when she and my great-grandmother Mamie were to serve tea to the mistress, Mamie accidentally spilled some hot coffee on one of the guests. In fury, the mistress spilled her own cup of hot coffee on Mamie. My grandmother reacted by throwing another cup on the mistress. She then grabbed her mother's hand and stormed out with her in tow. After that incident, Nana was determined to get her mother out of private domestic work. She actually managed to get her mother a job as a maid in a hotel. The hours were still long, the work grueling, the pay ridiculously low. Yet the "institutional" setting made both the work and the hours routine. There was no longer one "mistress" to whom Mamie was completely beholden.

But there were other stories closer to my own life that I was forced to confront for the first time. I realized I was implicated in the forces that had mistreated these strong and extraordinary women. It was more than my feminist politics and my attempt to hear these women's stories—stories that were so clearly about the class, racial, and ethnic divisions that divide women—that drew me to them. The phrases *the other mother* and *the second mother* kept running through my head.

In April 1993, I adopted a female child from Paraguay. She was six months old when I adopted her and seven months old when I brought her back to the United States. She was extremely sick when I first saw her. The first few days she was with me were spent in panic that she might not even survive the trip to the United States. Two women stood by my side and helped me get through the adoption process. Mary Elizabeth Bartholomew was in her third year of law school and approaching exams. She attended the City University of New York, and fortunately had generous professors who allowed her to complete her work at a later date. The other woman was Graciela Abelin-Sas, who spoke to me over the phone constantly. Graciela was born in Argentina and knew friends in Paraguay. These friends and contacts provided me with translators and access to the best medical care. An American Embassy doctor insisted that Sarita had to have an operation or she could not go to the United States with me. I was terrified and completely distraught. I had to pay him in cash for the operation. The next day, however, he signed the papers, after which there was no further talk about an operation. Sarita was put in the hands of an excellent pediatrician who had been exiled from Paraguay for political reasons and came back to vote.

For all those years, Paraguay had been under the control of the brutal dictator Stroesner. Like so many other Latin American dictators, Stroesner had been trained in the Panama School by the United States. The Panama School specialized in training military rulers who could control the "unruly" people's movements that were continuously breaking out in South America. Any form of resistance led to imprisonment, torture, and/or exile. Resistance was defined with a broad brush. In Paraguay, psychoanalysis had been ruled illegal under the dictatorship. Many of the people who helped me had had to leave their country. Many of them were psychoanalysts like Graciela. I entered Paraguay about three weeks before the country was to have its first election in thirty years.

People were enjoying the openness before the election and the oppressive weight of the dictatorship being lifted off their back. There were demonstrations on a daily basis, and citizens of the United States were warned not to go out of their hotel unescorted. One evening, Mary Elizabeth went out to a rock concert with some of our Paraguayan friends. My world had become my baby, so I did not participate in the cultural and political festivities that preceded the election. I had the experience often described by adopting mothers.[45] Sarita seemed like a miracle to me. I could not believe that I was actually a mother, much less the mother of this magnificent baby. Even when she was very sick, she was a force to contend with. When Sarita was first handed to me, I was shaking so hard that I thought I would drop her, and Mary Elizabeth had to put her arms around me to hold me up. Sarita's hair was slicked down so she would look white. The first thing she did as I took her in my arms was pull her bow out of her hair. She stared me down, bow in her fist. Her hair started popping up all over. She wore a little pink dress. She has never worn a bow or a pink dress ever since. She still stares me down all the time. What particularly amazed me was that she was born a hundred years to the day of my grandmother. I could not help but feel overwhelmed by the dreamy idea that, in some way, my grandmother had been returned to me. My grandmother was the only person I had known who had ever been to Asunción. When I walked out on the hotel room's balcony, I felt that Sarita and I were destined to be together.

I went out on the balcony to watch the demonstrations. I have always been an activist and, in my own mind, a militant anti-imperialist. I wanted to understand the issues in the demonstrations. My heart sank as I slowly made sense of the chants. They were directed against the "Americans" in the hotel. I never identify myself as an "American." Indeed, by then, I knew enough about "South Americans" to realize that the designation "American" be-longed more to them than it does to us, since so many of us identify with our European origin. The appropriation of the name "America" I took to be an exclusively imperialist act. But standing on that balcony with my daughter, I was positioned as an American. The signs against the adoption of Paraguayan babies that were being taken to the United States were addressed to me among many others.

Gayatri Spivak has frequently used the phrase "enabling violation." In her recent book, there is a footnote in which she suggests that adoptions from the global South by women from the United States and other parts of Europe are enabling violations.[46] I thought I knew what that phrase meant before I went to Paraguay. But its full meaning did not hit me until I saw my-self standing on that balcony with Sarita in my arms.

WHAT IS AN ENABLING VIOLATION?

As is frequently the case in her writing, Spivak finds "the words to say" what is so difficult to speak by putting words together that do not seem to fit. How can something be an *enabling violation?* Spivak wants to help us understand why we must always take a critical position whenever we think we have engaged in an ethical action in this thoroughly unjust world. She calls us to take up an endless critical stance in our attempts to be ethical—"us" is here broadly meant to indicate academics and intellectuals from the "first world."[47]

An action can seem enabling from one perspective and a violation from another, and these two perspectives can both claim some legitimacy. So what happened to me when I was hit with the meaning of her phrase? From the perspective of the individual future of Sarita, adoption enabled her future. She might not have survived had she remained under institutional care. She needed a great deal of medical attention to recover her health. I had scrupulously tried to follow the procedures established by Paraguayan law, including the provision that the birth mother be given four months to change her mind after signing her baby into adoption. I knew that Americans desperate for babies often ignored these procedures. I believed the stories about babies being stolen.

On the one hand, Sarita's adoption enabled her life. On the other hand, the demonstrators were right. It was a violation of the Paraguayan people to have their children sold or even legally handed over to the United States. There was also no doubt that babies had become a major export from Paraguay to the United States. I agreed when, in its first official act, the elected government prohibited any further adoptions to the United States and other countries.

My first attempt to take responsibility for the enabling violation of the global South was to completely let it shake up all my own ideas—ideas I did not even realize I had regarding kinship, citizenship, and language. I decided that my daughter must not lose her cultural heritage and I have taken steps to preserve it for her. I insist on her joint citizenship with Paraguay and the United States and will continue to do so.[48] The legal battle I could handle. But I could not keep Spanish or the native language of Paraguay alive for my daughter, since I could not speak either one. I wanted to educate myself. I was determined at least to learn Spanish. I wanted to learn about the cultural heritage of Paraguay. I wanted to know as much as I could about Sarita's birth mother and the forces that had led her to relinquish Sarita. I knew I needed help for this project. It was not enough to have Graciela, who came from Argentina, a nearby country, as a godmother.

I needed help on a much more mundane level. At the time I adopted Sarita, I was a professor at The Benjamin Cardozo School of Law. I had to work. So I sought out a nanny. Alicia was a countrywoman of Sarita's to whom Graciela introduced me in another one of those coincidences I tend to identify with destiny. She had been a teacher and an accountant in Paraguay. When she came here, she was unable to work in either profession because she did not have papers. She had no choice but to work as a nanny. One employer after another had promised to help her, but they never came through. I promised I would. Having been a union organizer, I was determined to be a fair employer: I was making things up as I went along. When I hired Alicia, I was hoping to find a "nannies'" cooperative so that I could do the right thing. All I found were agencies. So I tried to decide what I thought was fair. As I have now learned from the women of *Unity*, there was nothing I alone could do to make Alicia's situation fair.

Alicia was an invaluable source of support for me and for Sarita, during her first three and a half years. When I brought Sarita to the United States, she had parasites, several infections, projectile vomiting, and most seriously, water in her lungs. She often ran dangerously high fevers that, at times, caused convulsions. I remember sitting in the bathtub talking over the phone with the pediatrician who told me Sarita needed to stay in a cool bath until the fever came down. But it would not come down. Finally, in desperation, I called Alicia. She came immediately. Together we rushed Sarita to the hospital. She had scarlet fever and pneumonia. Alicia and I spent the night together sleeping by Sarita's side. Alicia did not live with us, but she stayed with us until Sarita was completely out of the crisis. This is but one story among many. I remain absolutely indebted to her—what I owe her I will never be able to repay.

Did I try, at least, to deliver on my promise to sponsor her? Yes, I tried. Alicia had contacted an immigration lawyer before she came to work with me. He lied to her outright about his strategy. He pushed her into a class action suit for an amnesty for which she did not qualify. She had signed papers without fully understanding what the lawyer was doing. In fact, he ensnared her in a set of lies concerning when and how she had come to the United States. I consulted a number of lawyers I trust, including my former student, Wendy Lazarr, now a noted and progressive immigration lawyer. Short of having Alicia divulge how she had been trapped into perjury, there was nothing that could be done legally to get her out of the case. Trapped or not, there seemed to be little legal hope for a new application.

I looked for a new solution for Alicia: marriage to a U.S. citizen. This was before the 1998 law that makes marriage to a citizen a much more difficult route to a green card. Even at that time, however, anyone who undertook to marry an "illegal" had to prepare himself or herself for the possibility that the INS would interrogate both parties about whether the marriage was in fact "real." Uncle M., my "adopted" brother, and an openly gay man, volunteered. He spent several months preparing. We even worked together on things like how to walk like a straight man, how to be sensitive to the woman he loved but not *too* sensitive. After all, he was supposed to be straight. So there were certain things he should not know, lest he give himself away. How much does the average straight man know about lipstick? Alicia worked hard from her own side. She learned all about what he liked to eat and read and memorized his poetry. Uncle M. is a talented poet. She found out all the basic facts about his life and education and committed them to memory. We found a place for them to live—no small feat in New York. We took romantic pictures of their courtship. We were two weeks from the wedding.

And then, for religious reasons, Alicia decided she could not go through with the marriage. Alicia is to this day a devout Catholic. The Catholic Church believes that it is a mortal sin to have a marriage of this sort. We stayed up all night trying to explore other options. There were none. Alicia was left with no choice but to continue living "illegally"—which prevented her from working in the professions in which she had been trained—or to return to Paraguay. She returned to Paraguay, where she now works as an accountant.

I miss her. There are pictures of her all over my house. Alicia was the ghost that haunted my first meetings with the women of *Unity*. Despite her absence, Alicia was the one who called me to the task of finding some solution *with* others to the unfair immigration laws of the United States.

We keep up. She will never be out of my life. She will never be out of Sarita's. But here, in this country, she did not get the life she sought. Like so many others, Alicia came to the United States because, under the dictatorship of her country, there was no attempt to sustain or maintain an educated middle class. After the dictatorship, some things were different. Alicia now does the work she was trained to do. My failure to help her changed me.

Behind her is another absent presence, another woman from Paraguay who relinquished her baby: Gabriela, Sarita's other mother. I have no doubt that someday Sarita will go and seek her out to hear her story—to discover what kind of relationship they might manage to have together.

LEARNING TO LISTEN: WHAT IS A DRAMATIC INTERVIEW?

I now realize that I came to *Unity* to make peace as best I could with those whose absences haunt me. There is a lot that women from outside the organization can do to support groups like *Unity*. Some of those ways are quite obvious. Since privatized childcare is not a fair solution, we must begin to fight again for decent publicly funded childcare. We can join the "friends" of such cooperatives and participate in activities like fundraising. We can offer our houses as training houses and make sure that, if we do decide to have our house cleaned by someone else, we try to do it through a cooperative. And in the name of Alicia and all the members of *Unity*, we can march with them until they are free people with all the rights we take for granted as citizens of the United States. But the first thing we need to do is learn to listen.

What did it mean for me to learn to listen to the women of *Unity*? My solution to the complicated problem of representation was to attempt to conduct "dramatic interviews" with the women in *Unity* who were bravely willing to engage with me—with all my positions and identifications—and let me record their words. Not having a green card and agreeing to talk about your life in an interview puts you at risk.

I have used interviews mainly in my work as a playwright. My idea of a dramatic interview is that I seek to let the person shine through in all her worth. Following Immanuel Kant, when we contemplate the beautiful we are able to escape the interest in actual objects that limits the free play of our faculties. An interest in an object is, for Kant, very broadly construed. We have an interest in an object when we seek to use it, bring it into being in a particular shape, or stamp it with value. When we conduct an interview or a survey with the hope of accumulating knowledge for a particular purpose, we have an interest in that interview. There is something specific we want to find out. We want to find out, for instance, the voting patterns of women who make over fifty thousand dollars a year. We know in advance the information we wish to gather about the people we interview. Sociologists, political scientists, and economists conduct such interviews all the time. They serve an important informational and educational purpose.

In a sense, a dramatic interview does not have a planned agenda. The goal of such an interview is not to achieve a particular kind of knowledge as part of a scholarly project. It does not have as its primary goal the retrieval of information about the person. It is instead meant to portray the worth of the person in all her stature rather than construe her value or analyze her position in a framework created by the interviewer. A dramatic interview attempts Kant's aesthetic disinterestedness. Disinterestedness in Kant means that we are freed from our day-to-day investments in the objects and people around us—investments curtailing the free play of the imagination, which might allow us to see people and things of this world differently. But there is another sense in which I am using a Kantian-inspired notion of human dignity as beyond any interest anyone might have in it. When we are interested in an object, we want to shape it, use it, value it, understand it, make it "ours" in some possessive sense of the word. The dignity of a human being, however, is priceless.[49] The first aspect of a dramatic interview is to let the dignity, and the reverence for its manifestation in any individual person, shine through.[50] If we behold someone in her dignity, then her suffering appears beautiful or sublime rather that pitiful.

It was Schiller, the playwright, who developed the dynamic relationship between the beautiful and the sublime in his notion of the *practical sublime*. For Schiller, epiphany is not possible without beholding the dignity of the character. Only if we have already been affected by the contemplation of her infinite worth can we find ourselves in awe of the character's struggle and identify with her sublimity. In Schiller's concept of the *practical sublime*, the beautiful and the sublime explicitly interact with one another. We find ourselves in awe of someone who succumbs to suffering, and we identify with that person in her moral struggle. This is why confrontation with that struggle, and the recognition of how dignity calls us to it, is sublime and not simply beautiful.[51] For Kant, the beautiful is the symbol of morality. But the moral law inspires awe in us when we feel its pull, either in ourselves, or in someone else. Confrontation with the dignity of others inspires within us the grave exaltation associated with the sublime. It took Schiller to draw out the implications of the interaction of the beautiful and the sublime in tragic drama. We must first contemplate the beauty of what shines through in the person—no matter what her actual station in life—before we can behold the sublimity of her battle with the force of events that threaten to bring her down.[52] Of course, I realize that I borrow Schiller's insight into the interaction between the sublime and the beautiful and place it into a context that is not a play with fictional characters, but rather an engagement with *living* people. The suffering, loss, and mourning the women describe and continue to endure are all too real. Yet these experiences function as representations within a certain dramatic scene, and I am inevitably implicated in its creation. It is precisely because they are real people that I am all the more ethically responsible for how I proceed in that scene—a place meant to facilitate the space for their self-representation.

A dramatic interview can allow the infinite worth of the person to shine through so that any adversity described moves us with the full force of the feeling of respect. But we can never know whether, if it is presented dramatically, the interview will have its desired impact on the reader or on the viewer. As Kant reminds us, a reflective judgment originates in our affect. If a dramatic interview works, it must work again and again as each new person is drawn in so as to evoke the *sensus communis aestheticus* of reflective judgment. The universal is "there" in our judgment and is shared if others respond to it, too. Kant's daring was to try to connect what is most individual—affect— with the universal that comes to us only through our reflective judgment. To feel it, to see the universal with your own eyes and ears and then share it with others, is to go through a process of translation as we communicate what we have undergone. Indeed, sometimes the universal contained in the judgment, "Yes, this is sublime," can only appear as such because we have undergone an experience of translatability. In the case of a cross-cultural or multicultural interview, the interviewer will rarely succeed in the dramatic rendering of the universal. She must be profoundly affected—torn apart, even—by her effort to hear what cannot easily be translated into her own language or translated through cultural signs familiar to her.

IMAGINING ME

This is a lofty enterprise. No wonder anyone who takes it on will so often falter. How often I have faltered! No matter how hard an interviewer tries to let those with whom she engages represent themselves, she is inevitably part of the scene. Not only do their self-representations come through her, but she is implicated in the scene through her own identifications and self-representations; how she is positioned by them, how she views them, and how they are viewed and interpreted by others. I made contact with the executive director of *The Workplace Project* through a mutual friend. As I wrote earlier, I was looking to make contact with what I thought was a "nannies'" cooperative. That was the first of many mistakes I made. The executive director, Nadia Marín, spoke to the women in the *Unity Cooperative* about my desire to meet with them and perhaps interview them if they would allow it. But she would not go beyond asking. Any other steps to encourage or discourage the members of *Unity* would compromise the respect for the organization's independence from *The Workplace Project*, which does not interfere in the decision-making processes of the organizations it promotes. Its board of trustees is entirely composed of immigrant workers. As a result, Nadia left it up to me to make direct contact with them.

My assistant and friend Constanza Morales-Mair called Mónica Díaz, administrative assistant at *Unity*, and explained my desire to talk with members about their organization and their political program. Constanza was raised in a bicultural, bilingual home—her mother is from Scotland, her father from Colombia—and carries dual Colombian and British citizenship. Mónica speaks English fluently, but she told Constanza that many of the leading organizers in *Unity* did not. Mónica agreed to speak to some of the members about whether they would want to have a meeting. The result of those conversations was that Constanza and I were graciously invited to a Mother's Day party sponsored by *Unity*. The members wanted to meet us before we went any further. In fact, they made no promises as to whether we would talk again after the party. I asked Constanza to be my interpreter because at that time I spoke little Spanish. What Spanish I spoke was mediocre or downright bad, depending on how generous the listener chose to be.

This made me remember that earlier that same month I had been on a panel with Gayatri Spivak. Spivak told an academic audience that talking *with* people from different cultures demands that those involved in the conversation be able to speak their language with mastery comparable to Spivak's own knowledge of French. Spivak became well known in the United States largely through her translation of Jacques Derrida's *Of Grammatology*. I appreciated the importance of Spivak's insistence that, if we are to speak to others in their language, we must cross linguistic borders. I had studied Spanish for years, struggling to learn how to speak it. After many classes, I decided to seek extra help. I began working on an individual basis with the sister of a close friend of mine who was visiting from Ecuador. I had proudly managed to translate, admittedly, with my teacher's help, *Spot Va a la Playa*, a children's story I found to be much richer and more complex than the English version I read to my daughter years before. But translating the *Spot Books*—I have worked my way through all of them now—does not approach the mastery of language demanded of one who translates *Of Grammatology*. My hope for linguistic competence was that, if I worked really hard, I might be able to make myself understood and comprehend what was being said to me.

Even though before the party I had begun to prepare a significant list of "chatty" questions, I knew that there was a chance—a good chance—that no one would understand what I was saying in Spanish. And I was right. Lousy linguist that I am, I am inclined to panic when anyone addresses me in another language. The result is that I also tend to lapse into German, which, besides English, is the only other language I know well. Rather than asking, "¿Cuántos hijos tienes?" I would ask "¿Cuántos *kinder* tienes?" When I was not incorporating German words into the conversation, I would use the wrong word altogether. For example, I would say *hijos* (children) instead of *años* (years). This was certainly not a "cool" question to ask at a first meeting. I kept trying and stumbling around through most of the party. That day Nadia introduced us to several of the leading members of *Unity*. I could so easily imagine how they imagined me: a middle-class *gringa* who spoke incomprehensible Spanish mixed with German, who could not dance and did not understand the instructions for the games. Everyone at the party was having a blast. The more I tried to join the fun by waving and calling out "¡Buenos días!" when I should have been saying "¡Buenas tardes!," the more I stuck out.

Without Constanza as translator of the interviews and as my interpreter at the party, I would have clearly been unable to explain my conception of what I wanted to do. Constanza speaks fluent Spanish. Constanza's knowledge of Spanish and her ability to share other important cultural signs allowed her to join the party, the dance, and the games in a way I could not. I was a complete outsider. A young white male intern of *The Workplace Project* finally approached me, with several members of *Unity* who had become intrigued by what Constanza had told them about my project. On behalf of *The Workplace Project*, he wanted to question me further about my work, and he translated our exchange for the women in *Unity*. The ideal of dignity, since it informed their own practice, had a resonance with them. They wanted to hear more about the way I intended to use it in my book.

The women agreed to a further meeting and consented to being interviewed under certain conditions. They wanted to have the final say on whether the interviews would be published. They wanted to have the right to rework their interviews until they were satisfied with them. The conditions were easy for me to meet since the whole point of these interviews was for them to be represented with full respect for their dignity. Since I am inevitably part of the scene during the interviews, I have left my botched Spanish in the transcriptions rather than try to edit and clean up my act. Constanza's interventions have also been included since at the beginning I had to rely almost entirely on her for both my questions and the women's answers.

I now turn to the stories.

Chapter V

STORIES OF UNITY

$\mathcal{T}$he first interviews with the women of *Unity* were conducted by Constanza Morales-Mair and myself during the months of May and June 2000. That summer, Constanza translated the three interviews that were given in Spanish. Dinh Tran, my assistant at the Rutgers Law School, transcribed the tapes of the two interviews that were given in English. The drafts were then handed over to the women of *Unity* for them to revise and add any necessary corrections. Antonieta chose to rewrite her own interview and Mónica Díaz, the administrative assistant of *Unity*, typed and translated the revised version into English. Some of the women have chosen to use their own names, and those who wished to have photographs in this book chose them. In November 2000 we had a celebration of the completed interviews. The stories that you will read here have been finally approved by the women of *Unity*.

INTERVIEWS

Zonia Villanueva
May 23, 2000
[English Translation]

Drucilla Cornell: [Zonia,] where are you from, originally?
Zonia Villanueva: I am from El Salvador, from the region of Santa Ana.
D.C.: What is your [full] name?
Z.V.: Zonia Villanueva. Villanueva is my married name, and Valle my maiden name.
D.C.: Zonia, why did you come to the United States, and when?
Z.V.: I left El Salvador to come to the United States in 1988, because the [war] situation made things very difficult for us, for our subsistence. Because we

did not have a job that would provide the necessary money for us to be able to live. So I decided to do so. My brothers were here. One of them. And I told him I wanted to come, and he told me to do so. So I traveled to the United States in 1988.

D.C.: Is your brother a U.S. citizen?

Z.V.: Yes, my brother is a U.S. citizen here.

D.C.: Zonia, are you now a U.S. citizen?

Constanza Morales-Mair: Have you managed to get your documents of citizenship?

Z.V.: I am in the process of doing so. I have a work permit for seven years. And now I have begun the process of applying for the citizenship. We will see what happens next.

C.M.M.: Zonia, you are applying for the residence, aren't you?

Z.V.: Yes, that's right.

C.M.M. [*to D.C.*]: She has done her paper work, and she's applying for the residence: for the green card.

D.C.: Did your brother file documents for you?

Z.V.: No, no. I have a work permit because of the PTC that [the] Immigration Office offered the refugees a while ago. It's a program—the PTC—that they came up with which allowed us to apply for a work permit.

C.M.M.: Zonia, what does PTC stand for?

Z.V.: I don't know. It was a document that we had to fill out to apply for the first work permit, and that was given to all Salvadorian people.

C.M.M.: For all Salvadorian refugees?

Z.V.: All Salvadorian refugees.

D.C.: Have you worked as a nanny in the United States?

Z.V.: Yes. I worked [as a nanny] in 1988. During the first few months, after arriving from El Salvador, I worked taking care of three children. I worked with a Salvadorian family, and then—although, really, they paid me very little—they made me feel good, because my time with them was good. About eight months. I worked taking care of their children: a five months-old baby girl, a year-and-a-half-old girl, and a three-year-old girl.

D.C.: Why did you leave that job?

Z.V.: I didn't leave the job. It was the lady herself who got me the next one. She was very fond of me, and she wanted me to earn more money. She found a job for me at a cosmetics factory. So it kept me from getting bored, and I liked it, so I . . . she told me to take care of the girls in the afternoon and go to the factory during the day. So that was my first job at a factory; a lipsticks factory.

D.C.: For how long did you have two jobs?

Z.V.: I lived right there in her home. I stayed with the girls when she went out. It was like . . . we trusted each other, you know? I worked with them like that for about . . . for about a year. And then my husband came from El Salvador. So I stopped working like that, and then, I moved here because I was in Suffolk County. And this, here, is Nassau County. So I moved from Suffolk to Nassau. After this, I continued working at a factory. I worked in a plastics factory and then I packed plastic bags in another factory. Later, well, I spoke to a lady and told her I liked working with the elderly. So she told me that was easy to do and that I should take a training course.

C.M.M.: Zonia, was working in two places too hard? Did you think that was a bad arrangement?

Z.V.: No, no. I think it was something she wanted to do so that I could adapt, since I had just arrived. She wanted me to adapt to the weather and everything. She wanted me to go out, and get some fresh air. I didn't consider it to be hard. It was hard [in a way] because I don't drive a car, and I moved around by bus, you see? And she—whenever she could—she would pick me up.

D.C.: Did your sister come with your husband?

Z.V.: My sister? No. My sister, well yes, a sister came. Her husband was also here, but there was also a brother that had come over, across the borderland, you know? and another brother, too. Three brothers and four sisters are here, in all. And now, well, my eldest brother is a U.S. citizen, and they—my brother and his wife—filed the documents for my parents who are here now.

D.C.: Is your sister always with you?

Z.V.: Not in the same house, but we do have a good relationship and keep in touch. Actually, they keep their own apartment, you see? But yes, we get together for certain celebrations, and we feel great with the family, now that Mamá has come and we're all together, as it should be.

C.M.M. [*to D.C.*]: What you asked was . . .

D.C. [*to C.M.M.*]: Was her daughter always with her?

C.M.M.: No. Actually, what you asked was whether her sister was always with her . . .

D.C.: Oh! Whoops! [*laughs*] Whoops! [I mean] your daughter, your daughter!

Z.V.: My daughter was born here. She was born in 1990. She is almost ten, and she tells me not to worry because she is a U.S. citizen and will not allow them to send us back to El Salvador.

D.C.: Your daughter was born here after your husband arrived?

Z.V.: Yes. I came over at the beginning of 1988, and my husband came in the winter of 1988.

C.M.M.: Almost a year later.

Z.V.: Yes. A year later.

C.M.M.: And your daughter was born almost two years after he arrived.

Z.V.: Yes, two years later.

D.C.: A beautiful daughter!

C.M.M.: What is her name?

Z.V.: Jennifer.

D.C.: How old is she?

Z.V.: Ten.

D.C.: My daughter is seven years old. Why did Zonia come to *The Workplace Project?*

Z.V.: Well . . . but . . . Should I tell you the story of my work with the elderly?

C.M.M [*to D.C.*]: Zonia had mentioned that she had also worked with the elderly. We haven't gone back to that. Do you want to know how she got that job?

D.C.: Oh! Yes!

C.M.M.: Okay, Zonia, please tell us about your work experience with the elderly and then about how you came to join *The Workplace Project*.

Z.V.: I, well, my dream was to get a better job and earn a little more. I have something [in me] that makes me say: "I can, I can," even if I can't, but it is my option to say, "I can do it." So I wanted to take the training course and take care of old folks, even though I didn't speak English that much. But I registered in an agency and got the certificate and all. Then I started taking care of old folks in 1992, until 1998. I would change agencies because, anyway, I wanted [to go to] where I could earn some more. So I took care of the elderly for about six years. Well, that was until there were changes made regarding the care of the elderly. Old folks were left without coverage, Medicare covered less hours, so it was hard for me to get a [full] eight- or ten-hours case a day. I would get three or four hours. To fill up eight hours a day, I would have to take care of five patients. Work with five patients . . . five patients just to have a full-time job. That's how I left the training school to work with the elderly. Then, in 1998, at the Church of [Our Lady] of Loretto, some flyers were handed out. They were an invitation, for those [of us] who were already housecleaning workers, or who were interested in getting that kind of job, to contact them [the organizers], you know? My sister had picked up one of the flyers, but I hadn't, and she said to me, "You know what? They called me. We can go if you want." She said that to me, and I said, "But I didn't fill in the form!" . . ."Ah! Let's go!" [she said]. I am always very curious and restless, so I went with her. Then, there was a man there that was from Long Island and he was saying . . .

C.M.M.: Excuse me, Zonia, let's stop here for a moment . . . Drucilla, are you following?

D.C.: A lot, I think. Zonia, keep going. Muy bien, muy bien.

Z.V.: So, that's how I stayed there, and the man from Long Island said that he wanted us to do something together—between them and the community of Loretto at Hempstead—something like a group, a group of housecleaners, or something like a children's day care. Something that would help us have jobs. It wasn't that there were jobs, you know? So in 1998, Nadia Marín was there [too] working as interpreter, in translations. She is the executive director of *The Workplace Project*. Well, there were many of us that had gone there thinking that there would be a job for us the next day. So that day was very hard, because people thought that she [Nadia] was going to tell them: "You're going to work tomorrow," and it wasn't like that. It was through a project, you see? Then . . . I remember they wrote down the names of those of us who had stayed in the meeting. Only about five of us stayed in that first meeting. So that's when Nadia said that here at *The Workplace Project*, they were also trying to help women find a good job, right? Well, that's what happened. About three or four months later, I received a message saying they had called me from *The Workplace Project*. So that's how I came to *The Workplace Project*, and then they told me that I was one of the persons that had attended the meeting at Loretto; that I wanted to help, and wanted them to help me find a good job. So, that's how we formed the *Unity* cooperative, [and how] four workshops were designed for us to have a little more respect and dignity in our workplaces, because that's what we wanted, you know? [We wanted] well . . . like a more permanent job and a better salary. That's why I came here, to *The Workplace Project*, and thank God, [through] the group we formed—it's been almost two years and a half—we have managed to have at least five houses for each of us. I feel that it is a dignifying job because I am the . . . I set my own schedule. It's three or four hours. It's sixty dollars and I go home, and I have time to pick up my daughter. I have time for whatever else I want to do at home. But at a factory, I would have to work eight hours a day for forty-five dollars. It's eight hours a day, and maybe [all day] on my feet. I [once] went to a factory—this is a very hard experience I had between 1998 and 1999—I went to work at a factory, and they put me to pull out small hot saws—hot iron—from those machines, with gloves and all, but I kept hurting myself. So—I'm diabetic—and I told the supervisor that I couldn't work because I already had many wounds in my fingers, and because of the dirt and all. I told him I was sick, and that I suffered from diabetes, and that I was worried about my wounds. So I asked him if there was any other position where I could do something else. And he said to me . . . He wanted to help me, because he said he had tried to speak to the manager and had asked him to transfer me to the packing section, but he told me that the manager had said there were no available positions there. Well, I worked only three days at that factory because I had to work with very tough material, iron. And now, well, I've been working with the *Unity* cooperative for two years because, although we don't vote, we have gained a little more respect. They don't put us on our knees. Because there are many houses where they make you kneel to scrub the floors, or clean and pick up after the dogs and all that. Now, at least, we don't do that. So . . . it's because of our own self-respect that we don't do it. We have been taught to value ourselves as working women that we are, with dignity and safety.

D.C.: I also worked in factories, and I have seen many accidents. A woman who worked next to me had her hand cut off by a machine.

Z.V.: Ay! My God!

D.C.: There was blood everywhere. We rushed her to the hospital, and the doctor said, "We can sew her hand back on, but you must find it. So the other workers and I went to the factory to get the hand. And it was crushed. The boss had thrown it into the garbage. So a worker held me by my feet in the garbage, and I looked for the hand . . .

Z.V.: Ay! Jesus!

D.C.: . . . and we got the hand. It was so badly crushed that there was nothing to do.

Z.V.: We hear lots of stories like that now, [they] happen to many people in the factories. Many immigrants tell stories like these. Both men and women suffer many accidents while working in factories. It's very dangerous. Many factories [do] have safety measures—chemicals and such—but in others . . . there's nothing. . . . Yes.

> *[During the translation we asked Zonia if the women in the pictures displayed on the office walls of* The Workplace Project *are the same five women who at-*

tended the first meeting. She points out that those pictures are not only of the first
women that were there; she added that nowadays there are a lot more.]

Z.V.: That picture was taken during another meeting here, when there were
more of us. See? But in these meetings, people would always do the same.
Fifty would show up and ten or fifteen would stay.

D.C.: How long ago was this?

Z.V.: Two years ago. From 1998 until today.

D.C.: Zonia, do you only work now through this program?

Z.V.: Yes. Only.

D.C.: Zonia, two questions. One: how do you ask for jobs.? Two: what is
your rights reform program—amnesty, for example?

Z.V.: Well . . . we give our support. That is, since we're already in *The Workplace*
Project, if a march for amnesty is going to take place, we're invited to participate, and we can go. Because, we also need it, right? This is for amnesty or if
our solidarity is needed . . . The last time, we went to Washington to ask for
the residence status. We marched. I have been supportive because I like to do
it, and I feel that if I can do it, some others will also be able to do it for others
that will come after. And now . . . about the job distribution that we have here
in *Unity*. First, we organized four workshops. Through them, they taught us
how to recognize which were, well, the reasons for organizing: What do we
achieve by organizing? What are the results when a group gets together to
organize? We learned to get to know each other. First, we learned to know
each other, as a group of women, here at *The Workplace Project*. We learned to
get to know each other through those four workshops; we learned about our
rights, and then we learned about the respect we deserve as working women.
And as to how we assign jobs to our fellow-workers, everything is based on
rules, you know? Because not everybody is responsible. We want the job, but
we have to be responsible. We did this through a set of attendance regulations: the woman who accomplished the four workshops was assigned to the
first house that was registered in *Unity*.

C.M.M. [*to D.C*]: The rule is that if you don't finish the four workshops, you
don't get the job, because it is also about learning to know the organization and . . .

　　[*to Z.V.*]: . . . and there are also workshops where you learn about
housecleaning, right? That is also included in the workshops?

Z.V.: Yes. That, too. The first four workshops we took were for us to get to
know each other, to learn and grow, to learn how to relate and speak to the
client, to learn how to organize a group of women and what an organized

group of women does. Those were the first four workshops. After those four workshops, we formed four committees, right? So it was a set of rules that worked out very nicely, because we learned a lot, and then we were taught about the committees that came out from those same four workshops: Regulations, Publicity, Education, and Finances. So each committee was in charge of its own [type of] work. For example, the Regulations Committee worked out a set of attendance rules, took note of the order in which the houses came in [were registered], who would be assigned to the first house, and why so? A set of regulations, you see? And Publicity was in charge of advertising: making flyers, distributing them, and publicizing for houses to register. Education was in charge of organizing theoretical workshops, and housecleaning workshops to train us on the kinds of liquids to use, what to take care of in a house, and how to relate to the client. There is a lady who is a volunteer that maybe you have already met. Her name is Nancy Ryan. She lends us her house for training.

D.C.: Which is why her house is always so clean!

C.M.M.: Yes, do you remember? She said her house was always very clean!

D.C.: So to get work through the cooperative, do you have to join *Unity?*

Z.V.: Yes. You must be a member of *Unity* cooperative, accomplish the four workshops—which I did—and belong to [participate in] a committee. Whichever you like. In that way, [we check] attendance, order of arrivals, punctuality, and that's the way we hand out the jobs.

D.C.: How many women are there in *Unity* now?

Z.V.: Approximately twenty-five.

D.C.: Muy bien!

Z.V.: This chart is the seventh round. It shows the order of attendance, the order of attendance to the committees [*Zonia points at the bulletin board where each person's turn and assigned houses are specified*]. So when the first house was handed out on February 3, 1999, that house was assigned to me. The very first one.

C.M.M.: You had accomplished the four workshops.

Z.V.: Yes. I was the first one to finish the four workshops, and kept coming back, and coming back, and showing up, and that's why they selected me as the first one [to get the first house].

C.M.M.: How many houses are registered?

Z.V.: There are approximately seventy houses. That changes because when the houses come in, they call to say that they need someone, and two or three weeks later, they cancel. Yes. Some come and some go.

D.C.: Do you only do cleaning or childcare, too?

Z.V.: Cleaning. We don't do childcare. For example, one of our fellow workers was told to clean the kitchen floor, which was very large, and they wanted her to get on her knees and scrub it with a brush, not a mop or anything. Like that, with the brush on the floor, crawling, you know? So she did this the first time, and then she asked the lady for a mop. Well, she didn't give her the mop or anything, so she told her that she couldn't continue working like that. So we called and canceled because what we want is for women not to get on their knees. Well, we always have good experiences and bad experiences, you know? We learn through each one of them.

D.C.: Zonia, are you feminists? Do you identify as feminists?

Z.V.: Yes.

D.C.: Where would you like to see feminism go? What is your feminism?

Z.V.: As a woman? I would like to achieve a success, you know? As a woman that I am, and as a feminist that I would be, job after job, I've enjoyed trying to see where it is that I can be more successful. I had the intention, once—because I like it a lot—of being a cosmetologist, hairstyling and all that. I tried, but I couldn't because I didn't have the legal residence. And yes, I would like to be able to study that, and get the license. That would be the first thing I'd do, because I'm very interested . . . I want to do it. And I don't lose hope that whenever the papers come out, I will be able to do it.

D.C.: Zonia, a last question. What are your dreams for your daughter?

Z.V.: My dreams for my daughter . . . [I'd like her] to study as much as she can so that she can become someone in the future, like a lawyer or a doctor. That's what she says she wants to be: a Latina physician, a pediatrician, that's what she says. Those are my dreams: to be able to give her the studies [education] according to my resources. For her to be able to be someone in the future. And not like me, cleaning houses and doing odd jobs. We need to study, because we come to these countries here . . . There are lots of opportunities here, but the situation as we arrive is very difficult because one comes here with that thing about wanting to work and wanting to make money to send over to our families.

D.C.: Thank you very much, Zonia. Now . . . Shall we eat?

Z.V.: Yes, but before, I also have a story about when I was taking care of old folks. They sent me over to the home of an African American, and when I got there, and she saw me, she said: "Oh, no! You're Hispanic!" and didn't let me in. So that killed me, because I started crying. I stood out there in that cold, you know? I stood with one foot in and one foot

out, and she said: "No! You're Hispanic! You don't understand!" She
said that I didn't understand, and she hadn't really spoken anything to
me. She didn't know whether I understood or not. So I said to her:
"Please let me in!" I said I wanted to go in—to call the agency, and let
them know that she was rejecting me. But in the meantime, I was crying
because I felt discriminated or, I don't know, I felt that something very
powerful had happened, and I was crying, and she saw me, but wouldn't
open the door. She kept it half-open. So I went back to the agency, and
I told them that if they were going to send me to take care of a case, to
tell the client they [the cooperative] had someone available who was
Latina, and [ask them] whether they wanted her or not. Because one
really feels terrible. Discriminated! And I said to her: "You have not
spoken to me to see if I understand."

D.C.: African Americans are discriminated against, and then sometimes they
discriminate. I worked as a union organizer and there was often a lot of
tension between Latinos and African Americans. It was a very difficult situation. We, the women of the union committee, worked very hard for solidarity within the movement, but it was very fragile.

Zoila Rodriguez
May 23, 2000
[Interview in English]

Drucilla Cornell: This is Zoila, Mónica's mom. Zoila, how did you come to
this country?

Zoila Rodriguez: How did I come this country? I've been legal . . . kind of
had to go through borders . . . There were so many people coming . . . We
were lucky that we could get a visa to come to this country. Now in 1973,
it was the first time that I came as an exchange student. Then I came twice
and went home and lived there for ten years. I came back for like a year
and half and went back home. And now we have ten years living in New
York. Now, it has been hard to get a job . . . a new life. For most of the
people, it's been a better life. For us, it's hard for so many reasons.

D.C.: In El Salvador, you were a secretary.

Z.R.: Yes.

D.C.: What did you work as when you first came here?

Z.R.: Housekeeper and baby-sitter.

D.C.: What were the hours?

Z.R.: Hmmm. From 7:00 A.M. to, you know, whatever hours, ten . . . eleven . . . twelve. Weekends. I had to. I was a driver for the kids, so I had to go for eleven, twelve, ten . . . to whenever they were getting off from the party, for the same salary.

D.C.: So they paid you a salary. No overtime.

Z.R.: Never.

D.C.: Did they give you vacation time?

Z.R.: They did. One week.

D.C.: Did they pay you for the vacation?

Z.R.: Yeah, they did.

D.C.: How would you describe your experience working as a "nanny"?

Z.R.: Well . . . it is very hard because what I thought was that the people were going to sponsor me. They did file the papers, but it took a long time because they never made any effort to follow up the procedures. I thought . . . my daughter . . . she is the most precious thing in my whole life—so if you're going to leave the kids, the most important thing in your life, to somebody you don't know, you should appreciate that with your life. I mean, if you leave the kids with someone you don't know, you should appreciate what they are doing for the kids, because you are leaving them the whole day, you know. Those kids . . . These people would go away for two weeks, and they were leaving me with their children, like I was their mother. The youngest boy used to say: "She is not my house-keeper, she is my second mother." This is how he introduced me to his friends: "This is Zoila, my second mom." You know, when I left I didn't have the guts to say that I am leaving. I couldn't. I said that I am going to my country and I will be back. I couldn't say goodbye to them. And I was feeling so bad when I had news, that he was too depressed when I left. I was feeling guilty you know, because I thought it was my fault. I don't know. So many things that came to my mind two years ago . . .

D.C.: So in a sense you leave your own children to raise other people's children and then it breaks your heart to leave them because you love them, too. Now, how do you find working at *The Workplace Project?*

Z.R.: Well, they've given me security. People need help, and they can do that: they can help. In 1990 I was working for some other people . . . people I worked for and that treated me like I was nobody. Like you don't have dignity. Like you don't deserve to be treated well. And one day, I was with friends on my free weekend and I told them: "One day I am going to have all my papers done, and I am going to be a legal citizen. One day I'm going to build some organization to help women like me. Help people who don't speak English and don't know how to defend themselves." I remember this as if it were yesterday, and it really happened ten years ago. But I didn't know in those days about the *Workplace Project*. They came out here, to the Latino community in Hempstead in 1992, and I was introduced to this place in 1996 for the first time. I went to a protest for a girl who had been fired. She was without pay for almost six months, or something like that. I realized: "There is someone else [in a] worse [situation] than me." I came to this place and started fighting for it, you know. And this place has taught me that we all deserve to be treated with dignity and that everybody has to respect that. We are asking for what we deserve already. That's what we're claiming. We deserve this. We deserve a better life: a life with dignity. You know. They showed me this here at *The Workplace Project*.

D.C.: That dignity is crucial.

Z.R.: Yes, because every human being is born with it.

D.C.: What are your dreams for your daughter and granddaughter now?

Z.R.: This is how I will answer, because there is a dream for everybody. It is a different dream for different people, you know? I have dreams that you already have only by being a U.S. citizen. I told them in a conference that my dream is the reality that you already have. I told them: "You were born in this country. You have something I want: I want to be an American citizen, to have your same rights. You already have that. That is my dream. Your reality seems to be impossible for me, because it makes me feel like I wasn't worthy of my own dream." For example, my granddaughter is six years old. She has been in every one of our protests since she was two years and half. One day, we were getting ready to go to a rally in Washington, D.C., and Alexandra asked me: "Grandma, how long do we have to protest?" I answered: "Until we can walk as free people in this country." Then, she said: "But, Grandma, I am free to walk in this country, and I know why." I asked why, and she said: "I was born in this country." Her mother—my daughter Mónica—and I have the same dream. Alex dreams of being a rock star, and she makes up her own songs. She has dreams of her own. My daughter's and my own dream is to support Alex any way we can for her to have hers.

D.C.: So basically, without amnesty, there is no dignity.

Z.R.: Basically we have been treated without dignity. You know, it's like you don't deserve anything good. That is the way I see it.

D.C.: Zoila, I just want to ask if anyone in your family is an American citizen.

Z.R.: Nobody.

D.C.: Nobody?

Z.R: I have a small family. I am the only child. Mónica is the only child. So we do not have relatives here. It was my mom and my uncle—her brother—who raised me. You know, as [his] own. And this [gestures to *The Workplace Project*], you know. I have nobody else here. This is like my family.

D.C.: Did you leave the country because of the war?

Z.R: Yes, I did. Yes, I did. It was the main reason. Because my daughter was a big girl . . . like, you know . . . she looked older. And she was afraid. I was afraid for her, also. Because we were living out of the city, and I had to travel everyday to take her to school, and this and that. So I did not want to take any risks. Anyway, so this is why I sent her first. She came first to Tennessee, you know, and after—three months later, I came over. Yes. But the war was the main reason.

D.C.: Are you a feminist?

Z.R: Yes, I am a feminist, but I am not an extremist. Because I like to be treated as a lady. If there is a heavy box that I have to lift at work or at home, I'm not stupid, I don't want to break my back! I am a feminist because I fight for the same rights that men have.

D.C.: You were telling me about how to raise a daughter.

Z.R: I was saying . . . When a girl comes to this world and a doctor takes her out of this, you know, the womb of the mom, and says, "This is a girl." I remember when I had my girl and the doctor told me, "It's a girl," I cried. And I didn't cry during labor. Even with all that pain, I did not cry. But I cried when he told me "It's a girl." There is a name for it: a hero. It's a hero! It's a hero. Because when a woman comes to this world she will have to live in this world and do so many things that a man will not do. Women will. Nobody else. So that was a hero . . . who came out of me. The first hero I met was my mother.

Mónica Díaz
May 23, 2000
[Interview in English]

Drucilla Cornell: Mónica, could you tell us again the story of how you came to be in the United States and what your mother's working situation was?

Mónica Díaz: I came when I was twelve, in 1989. I was an exchange student because of the war in El Salvador. And I was in Tennessee for nine months as an exchange student in a Christian school. When the war kind of slowed down, the organization that my mother was working for ran out of funds. So then she was out of a job in the middle of the war. She realized that things were not looking any better, so she decided to move to Long Island, New York—some place she'd never heard of—and come and try a new life over here. I met up with her here in 1990, April 1990. And that is when she started working as a live-in housekeeper, which also included being a driver and a baby-sitter for two kids in a pretty big house that she had to clean daily. So that left me . . . going to school and almost living on my own. Basically, with her guidance over the phone. The only time I got to spend with her was talking on the phone. She was promised residency. And we did get a lawyer and visited him twice. It was pretty expensive. But he did not really accomplish anything because immigration laws changed over time. Either he was not keeping up with immigration laws, or the bosses were really not paying him for what he was supposed to do. So they were not really paying for the residence fee.

D.C.: Did the lawyer achieve anything for you?

M.D.: The lawyer was too far away and not well located for her to have a good relationship with him, but she did get her working papers. Only for working. It does not make you a citizen or a permanent resident. It only allows you to work as a legal immigrant for a period of time. Every couple of months you have to renew it. So that's what she started as, and then I got pregnant when I was fifteen. I had my daughter when I was sixteen. I was already living on my own, and it made it hard that she was not there for me because she was working. She had to work day and night, either just watching the kids or having to clean during the day. And in '94, I believe, or '95, when a friend of hers spoke to her about *The Workplace Project* and how they had a women's group that met every Sunday, and you spoke about your problems in general, and how they had little workshops going on so you could, I guess, step out of reality for a little while, she started coming here and she took the workshops. There was a time when the kids she watched over took summer classes somewhere around Uniondale, I believe, and she used to come and take the classes for workers' rights and, then, she became a member of *The Workplace Project*. That is when she had the opportunity to apply for a position, and that is how she became a member of the staff at *The Workplace Project*. She was a secretary in El Salvador. And, you know, she was used to . . . She was not used to being a housekeeper, you know? I believe that her self-esteem was not all that great, and being that she could not be there for me, she felt guilty for whatever I did and whatever happened to me. So being that I did not get married, and I had a baby, and I finished school as soon as I could . . . I finished school in '94, and I was supposed to graduate in '95. Once she started to work here, it gave her self-respect back, I believe, because she was getting screamed at and being treated like nothing at her other job. And she had the chance to utilize the skills that she had learned from school and from other places in which she had worked, and now she is working here. And now we can spend more time together.

D.C.: You both work here, right?

M.D.: Yeah, now. The reason I am working here is because of the housecleaning cooperative, which is a branch of *The Workplace Project*. We have offices right next to each other. And we are much closer now. During the time she was not there for me—not because she did not want to, but because she couldn't—things weren't great between us. And I had the chance to meet my father for the first time and actually live with him, which was a good experience because he was not the best of a person altogether. And seeing that I had not met him before, he tried to be a father to me or at least to his extent. I was not used to having a father around. All I would give him was attitude, especially me being a teenager at that time. It was not the best relationship, so . . . not that anyone had spoken bad about him. Now at that time, I could see already for myself the kind of person he was. I remember that he had never been here for me, and that still he would charge my mother for me living with him. I mean, he would charge my mother for whatever he could, like . . . after school I wanted to take karate, and he'd charge her for that. He charged her for me living with him. It was only a month that I lived with him, but whenever she wanted to see me, and he brought me out from New Jersey—the toll fee and the gas and all that—she was getting charged for them. Just to see me. And . . . I mean . . . I was still his real daughter, and I have his last name, but he does not know how to acknowledge that.

D.C.: Is he a U.S. citizen?

M.D.: He has been a U.S. citizen since the '80s. He is a Gulf War veteran. He still serves in the army. He has his life, but no will to help me at all. Not even with my immigration status, neither with any economic things.

D.C.: So he did not apply for you, for U.S. citizenship?

M.D.: He promised just like everyone else. Immigration has promised. The place my mother worked at as a housekeeper promised. He promised. But he said the lawyer was too expensive and he couldn't do it. And now I am over twenty-one. I don't know if it is any easier now, and plus the last time I had seen him, in 1994, when he found out that I was pregnant, he didn't want me living with him. I haven't seen him now for so long that I really don't care. I'm still on my own right now, and I've been . . . I'm doing pretty okay without him, so . . . I'm not really good at begging. So. . . . Just take one day at a time. I already fulfilled the curiosity of wanting to meet my father, and it did not work. He could have helped me, so I need amnesty or something better for me to be able to go to school and better myself, because I graduated from high school when I was seventeen. I was seventeen when I graduated and was ready to go to college when the doors closed on me because I was not a legal resident. I was living here only as an illegal immigrant, which doesn't help me any. There is not a guarantee that I'm going to be able to stay here because I can be deported anytime

D.C.: Can your daughter be a U.S. citizen?

M.D.: My daughter became a U.S. citizen, because she was born here and her father is a citizen of the United States. So if anything should ever happen to me, he would be in control of her life. But he's not around; we didn't work out. Things didn't work out and so we separated in '96.

D.C.: Where is he now?

M.D.: He is in upstate New York, incarcerated. So right now, child support is not coming in and it is kind of tough for me . . . I have been on my own with my mother's help. I had been on my own with his help. I have never been alone completely. As of last year, I was like . . . to me it was something new, actually, like the first time that I had, you know, that I really found myself with nothing to eat at one point. I don't believe in welfare. But welfare doesn't really help that much anyway. Now that they have even stronger laws, that not even legal residents get decent benefits. And now that I am working here, I feel better because many bosses out there lower your standards so much and everything else, that they treat you like crap and it is really unnecessary. So right now, I am waiting for something better, for amnesty or for something, so I can go off to school.

[The phone rings. Mónica picks it up and takes a message.]

D.C.: Mónica, you're still talking about your future situation in the United States . . .

M.D.: My plan is not to go anywhere. I am staying in the United States for as long as I can. I mean, I have been here for over ten years, so my life is basically here. My culture is here. I mean, I am Salvadorian and I don't forget that. But I am a little bit Puerto Rican and I am a little bit black; I am a little bit American, and I am a little bit white. I am a little bit everything. I mean, it's not like if I am going to El Salvador and go into a Chinese restaurant. There are barely any. As I remember, I think there is only one and that one is in a high-class area, so it is not like I live here. It is not like here in New York, where there is something different in every corner. Everything is so different, and I don't think I would be able to accommodate myself to another culture because mine is basically this one. They always try to Americanize as much as they can, but nevertheless, it is still a third-world country. So I mean . . . I don't know what I would do over there. I would probably make a good secretary, because I know both languages, but . . .

D.C.: How did you become a secretary?

M.D.: I basically snuck into a free training school and it was a basic training school from Stony Brook. I took a little course . . . not over eighteen months. And they are free if you have a low-income family, so at the time—and I still am . . . I'm still a low-income family—I applied for it with my friend. She brought me there and they asked, just like any other school, for my green card, and I told them I had lost it, that I had to apply for another one and it was coming. And I stalled for three months, and the fourth month went by, and when a new semester started they realized that I had studied all semester without proof of a green card, and they insisted they needed it for the next semester. I couldn't go back because I didn't have it. So, well . . . I only learned a few things. I mean, whatever I did learn I am utilizing it now, and I am still learning, but I was taking the medical assistance course, and I only learned the office skills. I didn't get to the medical part because of my illegal status. That is always an obstacle for everything, especially for school, because, I mean, I can see their point: if you're not going to be here forever, why would they give you free school? But . . . if there was a way the government or the state could help . . . I mean, I have been here twelve years. That still counts as me living here. I guess they should get the point that I am not going anywhere, and that I now have a family here. So I don't know what the solution will be, but I'm still waiting for a legal residence and for anything to happen.

D.C.: Mónica, how do you live with the anxiety about the possibility of your being deported and your daughter as a citizen of the United States, and not of El Salvador, not being allowed to go with you?

M.D.: I try not to think about that because when I do, it really scares me because what happens is . . . that I would get deported, and I would live in El Salvador, and would probably make good money over there, but my daughter would be left here because she is not a citizen of El Salvador, and she wouldn't really be accepted because she is half African American, and the country discriminates against other races, especially the dark races. I mean, they love *gringos* that are white Americans. Once you are a little bit darker, and even if you are Salvadorian, you are not anything.

D.C.: Mónica, did you and your mother try to live together, and what was the neighborhood where your mother lived as a housekeeper and a nanny like?

M.D.: She lived in what I believe was mostly a Jewish neighborhood, and I mean all kids get cars when they turn sixteen, so it was a middle-class, rich neighborhood. So I wasn't really enthusiastic about moving into that kind of neighborhood, because . . . I mean . . . it was good and everything, but I didn't have the same financial background they did. So I was looked at as the housekeeper's daughter because I wasn't raised that way. Not that I think it was bad, but I wasn't raised that way and I was young at that time. And in El Salvador, I wasn't a housekeeper's daughter, I was an office worker's daughter, and for me it was a big change. It was a big change and there was a big difference between me and the other kids. My mother rented a room from people, members of church, that we had met in a mostly Puerto Rican and black neighborhood where I went to school. These people kind of took me in as their daughter, but you know, they were always aware that my mother was on the phone most of time, and tried to be as close to me as she could from a distance. She'd try to come to the house on Sundays and Mondays. So that was our living-together time. And it was like this for many years until we actually tried to move in together when I was eighteen. I was eighteen, and I thought I could rule the world! And I didn't need my mother then. So it didn't work out. Then I moved to New York City, so we kind of separated because she still worked in Hempstead, and I went to New York. But now we work together. It is a good experience, because we also see each other everyday.

D.C.: What are your dreams for your daughter?

M.D.: To be able to have almost—not everything—but most of the things I'd be able to give her. Things I didn't have when I was growing up. Like day camp and classes, and things like that that, right now, I can't actually give her because, I mean, the secretary part-time job does not pay the best, for a seventy-five-dollar-an-hour tutor and, you know, things like that. But

I'm trying to work for a day camp for the summer. We are in May right now. June is around the corner, and school is going to be over, so I am trying to give her that. I'm not trying to be extreme about rules. I'm trying to guide her to do the right thing. I teach her about bad and evil, and that things are good if you work for them, and that being good is the best way out. I mean, I don't hide anything from her. So whenever I have a problem she always knows it. She's not old. She's only six, but she understands. And she understands what I don't have, so she knows not to ask. Hmm . . . To me, she's a perfect child. Not only because she is my daughter, but everyone tells me so. I think that I am doing a good job so far. But there's always that financial thing that does matter, no matter how good a parent you are. So, you know, I would like to continue trying at school, and no matter what, make sure that I always have forty hours of work, and that I have a better pay and better benefits, and things like that for the future. I want her to always remember that I did whatever I had to so that we could be together, and that I was the one that provided for her with love and everything else.

D.C.: And did your mother leave her workplace with any benefits?

M.D.: No. The people that she worked for were factory owners in New York City and, unfortunately their factory somehow went down the drain and things didn't get any better. They had to sell the big house they had, and they lost their comfort and their one-million-dollar house. I mean, the structure of the house was only a million dollars. And they had to move from there to . . . back to the city where they were originally from. So they moved back there. Rent over there is not any cheaper, but they had to move. My mother said that they lived in a one-bedroom apartment. So I am sure that they don't have the best conditions right now, and they don't have it good since the time my mother left. They definitely didn't help her and didn't acknowledge the time that she spent with their kids as a "second mother." That's what the kids called her. The kids used to call her their second mom, and all the teachers and all the tutors knew my mother more than they did their actual mother. So she did not leave with any benefits, but you know, she had the knowledge of how abused housekeepers feel, because she experienced it more than once. Not only did she get screamed at, and whatever else, but the trauma of not being with me. Not just her, me also: I had to live alone for so many years and at such a young age, but that was like the best thing that could happen and that she could do at that time, I guess. So I take one day at a time now.

D.C.: How would you describe what working at *The Workplace Project* has meant to you?

M.D.: A closer encounter with my mother, and you know, it is an actual job. But it is also teaching me a lot about my life, and my background—since I have been so busy trying to fit in the United States—and a lot about certain people, and about what would be fair for workers and employers, and about all the things that immigrants have to go through. It's a good place where you come and let your feelings out or hear other people's feelings, because everyone is going through something all the time. Immigrants are always scared of being deported and having the INS call on them. They usually don't do anything about exploitation or anything like that because they are always so scared. So this is a good place that teaches you that you don't have to be scared for just being an immigrant. I mean, everyone else was at one time. It is funny that you know they were, but they don't recall it.

D.C. : Thank you very much, Mónica.

Antonieta
May 23, 2000
[English Translation]

Antonieta: My name is Antonieta. I am from El Salvador. I immigrated to this country in June of 1988. I came with my two sixteen- and seven-year-old sons.

I decided this because of the political problem in El Salvador, since living in a war is not easy at all. I'm from a city from the interior part of the country. To be more precise, from the east part of the country, where the armed conflict was much stronger, and where the encounters between the guerrilla and the military forces put the people that lived there at risk. And the guerrilla recruited the teenagers. To protect my sons: that was why I immigrated. We are still in this country, because even though the war is now over, there were still lots of problems there, such as delinquency and violence of all kinds; there is no respect for life, and here we feel more secure. And after twelve years of living here, we don't want to return. My sons are already used to this country. This does not mean that we have forgotten our country, even though the majority of our people is hard-working, respectful, and they want the best for the people, sometimes things take another turn, and the lack of work, the overpopulation, and teenagers that have been orphaned, children that grew up in the middle of the war;

all this has induced them to resolve their lives through violence. God willing, one day our country will recover its peace.

I am going to tell you a bit about myself. In El Salvador, I worked as a teacher. I gave classes in a high school, and I also worked as an administrator of a cooperative of small-scale coffee entrepreneurs. I traveled to this country with a tourist visa. Some of my relatives and members of my family live here, and with them, I enjoyed and learned how beautiful New York is. But since I couldn't keep on spending the money that I brought—my savings—I decided to look for work as a housekeeper, being that in any other type of job I would need to have an employment permit. Well, I began to search in the newspaper, and I read an advertisement that said: "Woman needed to sleep in, take care of two children and clean, wash, iron, cook, etc. English not required." I said, "This is ideal for me." So I made an appointment. They interviewed me, checked my identification papers, and gave me the job immediately. I told her that her house would be clean and—what was more important to me—that her children would be safe, physically and emotionally, because I was going to take care of them as if they were mine. Unfortunately the woman did not know how to guide her daughter. She would tell me "Don't let her go running from house to house around the neighborhood," and whenever I said to the girl that she couldn't go out for the time being because her little brother was sleeping or eating, and that I couldn't go with her, she would call her mother over the phone to complain that I would not let her go to her friend's house on her own. To which her mother would say, "Yes, go by yourself." This was a problem since, for this little seven-year-old, I seemed like an unpleasant person, and she would say offensive words to me, make signs and gestures with her fingers, etc. This made me quit the job because I was afraid something would happen to her on the street, and then I would be held responsible. It made me sad since I had grown fond of the two-year-old baby. I let the lady know why I quit and said that I hoped that in the future she would know when to say "no" to her little daughter. I say this because, if someone reads these lines, I want them to know that they can trust and rely on people, and that we give them a service, and that you have to live and protect your children but not say "yes" to everything they want, because when they are teenagers you want to correct them, and I think this is when conflict arises and the result is violence in the streets and schools. Don't use physical or emotional abuse to guide the children.

Well, after I left this job, someone I knew said to me, "There is a lady that needs a person to work as a live-in and take care of everything in the house. She pays one hundred and sixty dollars for five days, and she can give you the papers" [sponsorship to get legal papers in this country]. So I went to the interview and I started my new job the next day [September 2, 1988]. It was a family of four: the parents and a twelve-year-old boy and an eleven-year-old girl. They were very good people in the way they treated me. The job was to cook, and to keep the house clean and in order, and to wash and iron. I complied with all of my obligations. After a few months with them, I began to take the necessary steps to obtain my papers through a lawyer recommended by the lady. This made me spend one thousand two hundred dollars with the hope that I would become legal. I did it most of all for the sake of my sons' future. But the sad thing is that, by the end of 1994—six years later—my boss told me that she could not sign the papers because she didn't want to have any problems for having hired an undocumented alien. Then I felt abused, utilized, and tricked. Well, I had deprived myself from being close to my sons. If I had accepted a salary of a hundred and sixty dollars a week, it was because I felt grateful since I thought she was going to help me and that some day I would legalize my situation. Financially I wasn't well off. I had to pay rent, food, etc., and what's worse, my sons were alone in the apartment. I was with them only on weekends. The sacrifice was in vain, but I thank God for giving me good sons. They corresponded by behaving correctly: they did not wander about on their way from home to school and vice versa. The eldest worked and studied, and took care of his little brother. Can you imagine the worries that I went through, thinking about the dangers around them? Always praying so that nothing would happen to them, and that they would listen to my advice, being that I am divorced and that they didn't have a father close by. Imagine what could have happened to two minors on their own. So I had lost the opportunity and the time, since my brother could have asked for me as his direct family.

You asked me if I received social benefits. No. I never had any type of benefits. You ask me if I ever asked the people I worked for whether I could keep my sons with me. Truthfully, I never asked them, but yes, I knew beforehand that it wouldn't be possible. One knows what to expect of people. I will tell you something so you have an idea: the husband is a dentist, and one time I needed his services because I had a dental piece that was damaged. He took care of it, and then he charged me for his services. Another day I asked for a raise, and they told me they couldn't because they already gave me housing. I explained to them that for me that was not saving because I had to pay for an apartment so that my sons could live there, and I also asked them to allow me to go out every night with the promise that I would be there early every day, but they didn't accept and said that, in order for them to give me the papers as a "housekeeper," I had to sleep there.

D.C.: At what time did you have to be at work?

Antonieta: At 7 A.M.

D.C.: At what time would you have been able to leave?

Antonieta: At 8 or 9 P.M. After dinner and after cleaning up the kitchen. As you can see, it is not easy for immigrants to move ahead.

How did I get to know about *The Workplace Project?* One day I read an ad in the newspaper. It said that the project offered a course for labor organizers that consisted of seven two-hour classes. These classes were to learn about labor laws and organizing. It interested me because I would acquire some knowledge about our rights, and some day I would be able to take the message to another person. It helps us to avoid the abuse and also to fulfill our responsibilities. After I received my diploma—that was in 1993—I couldn't get too involved in the project's work because I still worked as a live-in. After I quit that job, I became an active volunteer. I have been a member of the board of directors since then.

D.C.: I am so moved. The world must change so that this doesn't happen.

Antonieta: I think that for that to happen, people who require our services must become more humane and conscious that, as human beings, we have rights.

C.M.M.: How was it that your brother obtained the citizenship and was able to request the residence status for you?

Antonieta: My brother came to this country more than thirty years ago, and my mother about twenty years ago or more. They became residents through an amnesty, and now my brother is an American citizen.

C.M.M.: Through your brother and mother, then, it would have been easier to file your residence documents.

Antonieta: I would have had them by now! After those lost years, I had to apply again, and my mother is sponsoring me. If we didn't do it like that to begin with, it is because I was advised that, through my bosses, it would be quicker because they were citizens and people with money that needed my services, and that in that category, there were more visas available.

C.M.M.: You were saying that working at the project and passing on the message to other women is a way of changing the world, and that this feeling has motivated you to continue . . .

Antonieta: Well. Going back to my work as a volunteer member of *The Workplace Project* and being in the board of directors . . . I found out that they were organizing a housecleaning cooperative. They told me to become a member of the cooperative because there would be jobs and we could struggle to get a better pay. The idea did not attract me because I was working six days a week, but since in El Salvador I had had experience in organizing, and also, later, in the administration of the cooperative of small-scale coffee enterprises, I had already lived through the process of achieving as a group what individually was not possible. I wanted to know if we could achieve the same here. That is how the cooperative came about, and I participated in the first workshops, and then I joined the Education Committee. This is when the most important part of being a member of the cooperative began for me. I started to be an active supporter, and I gave workshops to the new members. Let me clarify that the workshops are not only given as a requirement, but also to create a higher level of efficiency in our activities, because knowing the reason for the existence of the cooperative gives us security and confidence to do a good job, and to develop as people. As you can see, education about the cooperative is a previous requirement and also a permanent condition. Until now, everything looks good. This is what our cooperative is, and if my case moved you, Drucilla, it moved me even more to know about cases of women that have been exposed to real abuse and humiliations in their jobs, and that they were not only abused by the boss, but by the agencies that are dedicated to the housecleaning business and that exploit them without compassion. This is the difference between the cooperative and such businesses that are only interested in the money. They profit when they find a job for a woman, and they don't care how they are treated at the workplace. Our cooperative is interested in having the woman treated with dignity. That's why she is taught to learn about her rights immediately, independent from whether she stays or not as a member of the cooperative. What matters is that she receives the message. This makes us feel that what we are doing is worthwhile, and it motivates us to go on, because little by little, the world is changing. This is what motivates me to continue with the cooperative.

I don't want to omit that I have one client of the cooperative—one house—and for me, they are part of my family. They have my *cariño*, and I have theirs. It is not only a service in exchange for money. I am happy working for them and I participate in their joy and in their sadness. And being a member of the project is a great experience. It is a school. We learn that organized, we will improve our quality of life. It is a daily challenge and it is where all immigrant workers find support.

D.C.: Are you a feminist?

Antonieta: No.

D.C.: Why? Is there anything bad about being a feminist?

Antonieta: No. I am a woman that has to work because of necessity and not to compete with men, but because we have the same rights. I believe we are the same, or more intelligent than them.

C.M.M.: What do you understand by *feminist?* What is your definition of feminism and of a feminist person?

Antonieta: Well, feminism is that we have the same rights as men. But we should not think that men are our enemies, nor take it as a duel. I need to work and I will do it to prove that I can take care of myself. Preferably, a woman should be dedicated to the home. It is the hardest and best role that she can take on; serve the community through the church, the Red Cross, etc. There is no doubt that I am confused about this. For example, I have heard some of my women friends say, "If my husband or boyfriend goes to a nightclub with his friends, I have the right to do the same thing," and there I disagree, or it is just my way of thinking. Men and women, we have the same rights, but not the same privileges.

C.M.M.: So you do not identify as a feminist?

Antonieta: No [*giggles*]. I think that I do not identify as such, you know? Just as a struggling woman that has two sons to support. I am head of the family like many women in this country. The truth is that I don't know what feminism is, and it would be interesting to learn about it.

D.C.: What is your vision of the cooperative *Unity* in the future?

Antonieta: My dream is that there will be more members, that it will become strong and that we are one hundred or more. But I mean this, for its growth, not just to boast about it: "Oh! *Unity* cooperative of *The Workplace Project* has a hundred members!" No, but because these women will be making the difference. Because we do not only learn how to defend ourselves, we also learn to accomplish our obligations. Because if we demand our rights, we have to fulfill our responsibilities. Only by being united we will manage to improve our quality of life as immigrant women. The cooperative's growth would be fantastic. Imagine, each one of us taking the message to someone that for a reason or other is not a member. At least, we can let them know that they do not have to allow the abuse, that we have to look for a solution, that they do not have to let their dignity be stepped upon.

D.C.: Is the *Unity* cooperative part of *The Workplace Project?*

Antonieta: Yes, it is part of the project in the sense that it is there where the project was born. They organized us; they keep on giving us their support, their counseling and advice, and most of us are members of the project, which means that we are part of this big nonprofit organization.

D.C.: One last question: Is there anything that women who are not in the cooperative could do to offer solidarity?

Antonieta: Well, just by the fact of being here, you are already giving us your support. Because you are interested in the project, you're interested in helping the women of lower resources, since it is this condition—the need—that makes us allow the abuses. If you let others know that we are human beings and that we deserve respect, this will also help us raise consciousness among the bosses.

D.C.: Yes.

Antonieta: Because they are starting to realize that in the same way that they, or you, work on a schedule, we too have the right to a schedule that will allow us to live in a dignified manner.

D.C.: Antonieta, thanks a lot!

Antonieta: From the heart, the cooperative and myself, we thank you for your support in our activities.

M. E. A.
June 8, 2000
[English Translation]

Drucilla Cornell: When did you arrive in the United States and where are you from?

M.E.A.: I arrived in the United States in 1988. I've been here for almost twelve years. I am from Ecuador. I worked, the first time, as a live-in, a domestic worker. But I also worked in a factory for ten years. But about two years ago, the company was about to shut down, and it didn't shut down and it didn't . . . until finally they sold it, and I discovered the cooperative was better. Because, since I got to know the cooperative, I got involved and started working with it. And always looking out for a better schedule, because I am mother of a family. I have three children. I needed time for my family, and not spend so many hours working at a factory. In a factory you have to put in many hours. Right now, I work four hours and I leave early to go pick up my kids at school. I have time to be with them.

D.C.: Were you in charge of children?

M.E.A.: No, because there weren't any children. But when I went to work for the first time, there was a lot to do around the house. That house had not been cleaned in years! So when she saw that everything was clean, she brought her mother in. I had to take care of her mom. Besides cleaning the house, she would take me to clean up the office, and once the office was done, nice and clean, I helped her with the filing. I did whatever she told me to: washing, ironing, cleaning up the house . . .

C.M.M.: How long did you stay with this lady?

M.E.A.: Almost eleven months. She would also take me to clean up her daughter's apartment . . . and they never gave me extra money. Always the same.

D.C.: Would you mind if I asked how much you were paid?

M.E.A.: They paid me $100 a week.

D.C.: Did you at any time get a raise?

M.E.A.: No. When I asked to work more, to make more money because . . . I asked whether she could give me a reference for me to be able to work with someone else. She said to me: "You can work elsewhere now because I don't need you here anymore." She sent her mom back to Guatemala, and she told me I could look for another job.

C.M.M.: So she was from Guatemala, and she let you go?

M.E.A.: Yes, and she also asked me not to tell that I had worked with them because her son is a lawyer; to not damage his reputation.

C.M.M.: Was that a threat?

M.E.A.: I didn't take it as a threat, but rather that it would be detrimental to her because of her son being a lawyer.

D.C.: Which is a threat! Because how could you have a reference?

M.E.A.: I think it was ignorance because . . . if someone asked for a reference, I couldn't provide it.

D.C.: Right!

M.E.A.: Then, a nun from the church of Our Lady of Loretto referred me to a job with another lady. She was American. They treated me a hundred percent better than my previous boss, because this one was very considerate, and she even paid better. Fifty dollars more, which was very good for me.

D.C.: A hundred and fifty dollars.

M.E.A.: And work there was not even a fourth of what I had to do at the other place!

D.C.: Did you take care of kids there?

M.E.A.: They did have children there, but I took care of them only in the afternoons. They went to school in the morning. I only had to make breakfast and dinner.

D.C.: How long did you work there?

M.E.A.: Five or six months. The lady was pregnant and a baby girl was born. So I asked her if she could refer me to somebody else, but she couldn't. Because she wanted to send me to a friend of hers, but then she told me: "If anyone asks for a reference, and you find a job . . ." to give her phone number and her name and . . . that's what I did when I found a job. But she wanted me to work during the day, and then go back at night to stay with her so that I wouldn't have to pay any rent.

C.M.M.: To help her with the children?

M.E.A.: No! So that I wouldn't have to pay any rent. Only to sleep, she said. But I found a job somewhere else, but as a live-in.

D.C.: You didn't have to work?

M.E.A.: No. She said, "No. Don't worry about dinner. I'm here and I will take care of everything at home . . ." I should go there only to sleep.

C.M.M.: This new job would be, then, your third job . . .

M.E.A.: Yes. This was another job that I took as a live-in. The children there were older.

C.M.M.: How many children did your previous boss have?

M.E.A.: Three.

C.M.M.: And four with the baby?

M.E.A.: The baby girl was the fourth child.

C.M.M.: So you took care of the three oldest children?

M.E.A.: Yes. And when I arrived to the other place, which she had referred me to, and where I was very well paid. What didn't work out was that I decided to take a medical test, and they found out I had something that needed to be treated, so they prescribed me some medicine. So the lady didn't take this well. She thought it would be contagious . . . that she would catch it and so on . . . And she told me to leave and come back for the job after I got better. I had a bad cough. I sometimes had to work outdoors, in the sun . . . go fetch the horses. And inside, there was too much air conditioning. I don't know. I caught this cough. . . .

C.M.M.: Working in the countryside?

M.E.A.: No, at a mansion in Hicksville, New York, rather far from here. They had horses. I had to go, take the horses to where they ran and ate. There . . . in the fields. I had to take them in the morning, and bring them back at three in the afternoon. But indoors, there was too much air conditioning, and I caught such a bad cough! And she said to me that she might catch it, and that I should go see the doctor and have a checkup and then come back with a letter from the doctor after I got better.

D.C.: Did she pay sick leave while you were sick?

M.E.A.: No. I had just been there for about three weeks or fifteen days. I don't remember. She paid everything [she owed me], but I found a factory, and in that factory I stayed on for ten years, and . . . I . . . didn't want to think about it.

D.C. [*to C.M.M.*]: M.E.A. had bronchitis and this woman was paranoid?

C.M.M. [*to D.C.*]: She was worried about what M.E.A. might have.

D.C.: You know what I think? I think she thought you might have TB—tuberculosis. Sometimes racism takes the form of "You have a contagious disease."

M.E.A.: Yes. When the doctor gave me the medicine, he asked me: "Are you going back to work there?" I said no, and he said: "Okay! Do you have a job?" I said yes. That was when I went to visit without his letter, the boss sent me back to get it. When I went back to the doctor's for my checkup, he said: "But it was just a cough! What a fuss!"

C.M.M.: A cold!

M.E.A.: Only that the air conditioning was too high and it was bad for me. Well, I was better off working at a factory.

D.C.: What kind of factory?

M.E.A.: It's a factory of . . . It's a warehouse.

C.M.M.: How much did you earn there?

M.E.A.: I had an entrance salary of six dollars the hour. There was a union and all, but since they never raised the wages, I—with the classes that I took here [at *The Workplace Project*] and that taught us about our rights [and] that we had to claim them—I went and presented my claim. But we picked up signatures and they gave us a quarter. Twenty-five cents!

C.M.M.: Was that the raise?

M.E.A.: That was the raise. After eight years!

C.M.M.: How many hours a day?

M.E.A.: Eight hours a day. I put in forty hours a week. . . . Sometimes I'd work overtime, which paid time and a half.

C.M.M.: After how long did you ask for a raise?

M.E.A.: I had been working there for eight years.

D.C.: Oh! No raise!

M.E.A.: No raise.

D.C.: What were your benefits?

M.E.A.: We had holidays; we had vacations and insurance. I had no family coverage. But what I didn't like was that the union didn't do anything . . . that they didn't pay attention to any complaint we brought up, but when they—the Union—were going to charge us more, they would do it without notifying us at all. They just did it. For example, the union would charge us more without letting us know. They just took it!

D.C.: Did you have health benefits?

M.E.A.: We had six sick days, and they took one. They left us five, and they didn't even tell us about it. In other words, when they were going to negotiate a new contract with the workers, no one would speak to us or say: "This is what we're going to do" or "This is what the company says we're going to do." Only when it was done . . .

D.C.: I worked in a union, and I was fired because I told the workers that they were not getting the benefits the union told them they were receiving. A corrupt union. What was the name of your union?

M.E.A.: I don't remember which one it was. When I went to the union to complain about them not talking to us when our sick days where going to be taken from us, he—the union director—said that he was our representative, and that as such he was authorized to make all these changes. "Well," I said, " if you represent me, you should also represent us by getting us a raise. What you represent is their asking us for a higher production." Because yes, the production goals they were demanding from us were high, but a raise? Never! They didn't give us a raise!

D.C.: When did you first start working with the women of the cooperative?

M.E.A.: I was working both at the factory and with the women of the cooperative. I've been working two years here. In other words, I was working weekends at the cooperative. Saturdays. Because we have the meetings here on Sunday, and I went to the factory Monday through Friday. We meet here on Sunday, once a month. On Sunday, I came to the workshops.

D.C.: What made you come to the cooperative?

M.E.A.: I took a course on labor rights. When I took it, then I started my
work at the factory, too, helping the other women workers to learn their
rights: that we had to complain about the union; for them to not treat us
the way they did—we were yelled at: "Spanish . . . shit"—by the boss . . .
Because we would ask him why he made us fill out orders that were very
heavy. We would fill in the orders, but we asked for a man to pick up those
that were very heavy. And that's when the boss would use that kind of dirty
language! You see? We, the women, picked up the boxes and we organized
them. For example, there was this sheet with a list of orders that had to be
taken out: perfumes, medicine . . . whatever the client had ordered, and we
had to pick all that up. And all that was too heavy because there would be
pots and pans, and heavy metal items . . . chairs . . . Too heavy! We were
asking for a man to be brought in, and the boss didn't want that. So there
was a young girl that confronted him. What happened was that this girl
did not put in the heavy items, so the boss came up to her and took the labels from her, and said: "Go home!" So since I had more or less learned
about workers' rights, and I knew that he was being unfair to us . . . I had
told other fellow workers that we had the right to stand up to him. So the
next time another incident happened, we refused to work a full day.

D.C.: Do you consider yourself a feminist?

M.E.A.: Yes. I am a feminist.

C.M.M.: Could you tell us more about what you mentioned about how you
were concerned about the well-being of the older women workers?

M.E.A.: Well, there were ladies there that were forty and fifty years old
and who had been working there for about twenty or twenty-four years.
Well, I'd been there for ten years, and I felt young, but my energies were
drained out. I'd get there tense and nervous. I couldn't work hard. So I
said to them: "How come they keep asking us for higher levels of production when our energies are already drained out? Those women
couldn't work any harder! And also, there were nineteen young women
that were on a list that the boss had and that he was going to take into
the office for them to sign some document, even if they didn't want to
do so. But the supervisor said that they had to sign and that if they refused, the supervisor would do it [forge it], stating that the worker had
signed it. And that was not fair. Because I admitted that I couldn't work
any harder. Even if I tried, I couldn't do it. And they would force the
older ladies to do so. That's when we started mobilizing. Because that
was too much exploitation.

D.C.: [*to M.E.A.*]: one more question. What do you expect for the girl . . . for your daughter?

M.E.A.: Not only for my daughter. I would want it for other children too. Actually, I want them to study, to get educated. Because if one has a title in something, one can work wherever one wishes. Because unfortunately I didn't have that opportunity: my parents were very poor. And I seek that for my children. Something. A better future so that they can make a life for themselves.

D.C.: One more . . . One more question! From where do you get your courage? I know how hard it is to fight the boss!

M.E.A.: Yes. Maybe for being such a fighter, and for claiming my rights, I have lost other things. Right now, I'm needing a letter that has to be filled out for the green card, because that's what Immigration requires. I asked for it at work. Having done all that, they told me that they would send it to me by mail, and till this day . . . I have not received it. They said that the reason for their not giving me the letter was that I had confronted the boss and the union. I wasn't on my own. I went with another fellow worker, and the others saw us. There were two of us. I haven't received the letter. I needed that letter. I think that they felt I was doing something for justice . . . for the workers to be respected. If they don't send it, well . . . I'll get it somewhere else. I'm now going—maybe this week before they close—to where the main offices of the company are located, in Westbury, New York, to ask the lady in charge of all the workers for the letter. Because I asked the union for it, and they still have not sent it. I asked the supervisor there, at the factory, and he said the union would send it. Because I answered back. I have been told, by my supervisor, that [it was because] I have an attitude and had asked: "Why do you demand such a high production? I am not a machine to just throw it out there! I can't produce miracles!" That woman had [already] presented some complaints about me, saying that I didn't want to do my job. That time they even went as far as firing me. But I had other supervisors [that were] on my side, and I went back in through another department, and went on working with others, until she noticed I was back at work, in spite of having been fired by her. All of this for claiming our rights. I don't work for that company anymore. I am very happy working with the *Unity* housecleaning cooperative because I have more time to spend with my children. I am learning English, and many other things.

Chapter VI

ABIDING BY LAST RITES

> *Here is the strong one, the other*   one
> *Here is the strong one, the other*   *after the other*
>
> *Here is the strong one, the other*   *after the other*
> *Here is the strong one, the other*   *after*
>
> —*Alicia Ostriker, "Mother/Child"*

> *What can I possess*
> *But the history that possesses me.*
>
> —*Alicia Ostriker, "A Meditation in Seven Days"*

*O*n the last day of her life, my grandmother called me at ten in the morning. The purpose of the call was to review with me the last rites she wished to have carried out. She explained that her dead husband's spirit was waiting for her at the end of the bed and that she would be leaving with him that evening. So we had to go over the rites right then. After she died, I was to take her rings off her fingers, put them on mine, and then put the rings that she had given me years before on her fingers. She also wanted my first three published articles to be buried with her. At her sister's insistence, when she was in her eighties, she had an audience with the pope and received a paper that exculpated her from the sin of having married a divorced man. She asked me to put the document under my articles: "God will have to read you before he gets to the decision of the Catholic Church." We went over all the details of her funeral: the kind of coffin she wanted, what she was to wear, and how I was to perfume the petals that were to be gently scattered over her body. I was to give the eulogy and make it as powerful as one of Eva Perón's speeches. She wanted her favorite Broadway tunes playing at the moment when the mourners started to gather so that they would not be sad. She entrusted my mother and me to choose the right tunes for the tape. The service was to be held outside the mausoleum she had built for her husband when he died, and where she, too, was to be buried. She wanted to make it clear that this was not our last goodbye. She promised to visit me regularly and asked me to greet her spirit every year on the day of her death. She had set an appointment with the manicurist, and got off the phone as soon as she arrived: no one in my family dies with a chipped nail. Later that afternoon, she put on her makeup. We do not die without makeup either. She had a nurse the last months before she died. She told her she would not be having a meal that evening because she would be going to Warren before dinnertime. The nurse did not pay attention to her and made dinner anyway. At 5:43 P.M., with perfectly done nails and makeup, my grandmother suddenly sat up in her bed and reached out in front of her. When she fell back, she was gone. The nurse was so surprised that she dropped the dinner tray. The doctor said she did not really die of anything, just old age. On the death certificate, he wrote that she died of the flu. I carried out the last rites as my grandmother had told me. Since then I have always worn her rings, and for the eulogy, I tried to give as stirring a speech as I could.

My grandmother had to be laid out so that I could personally take her rings off, put mine on her fingers, and put the reading material in the coffin in the order she had requested. I was not quite sure how to go about finding someone to do a ceremony so as to greet her spirit on the anniversary of her death. I knew a woman I thought might be able to help, who had been one of my closest friends in our consciousness-raising group, *Las Greñudas*.[1] Her name was Muriel. During the years of our time together in the consciousness-raising group, one of the tragedies that struck the whole group was the death of Muriel's mother, Miriam. Miriam had been a longtime activist and fighter for justice. She had been in the Communist party and one of the organizers for the union 1199 at the Harlem Hospital, where she worked for many years. Miriam had also been an activist in the civil rights movement. The vigil after her death was run not only by a minister, but also by another woman, Henriette, a longtime friend of Miriam's. I thought that some of the rituals and prayers in which Henriette led the mourners were not Christian. I was impressed by the beauty of the rituals, but too sad to question Muriel about them. However, Muriel thought that Henriette might be able to help me. I knew that in her day job Henriette ran a community center that helped drug addicts undergo detoxification. I approached her shyly, knowing how busy she was. She told me that she had never acted as a priestess for a white woman before, but trusting the sincerity of my request, she agreed to try.

One of the conditions of these sessions, at least as I was taught, is that the priestess knows nothing about the person's secular life, has never seen a picture of her, has never heard stories about her, and has never even known anyone close to her other than the individual requesting the service. A person who requests such a service has to prepare for it a whole day in advance. The priestess gives the recipient the most magnificent of gifts: the visitation of a beloved soul that has passed beyond the secular world. These are not the kind of séances portrayed by movies, the media, and popular culture, in which someone moves objects to the accompaniment of weird noises. Nana always warned me against those phony séances.

For the next ten years, I religiously greeted my grandmother's spirit on the day and time of her death. I missed her too much to ever pull myself together to ask questions the way my grandmother did of Warren Kellow. Although the spirit was practical, as was typical of my grandmother, the sound of that beloved voice would send me into a grateful reverie. Once finding me in great sadness over a man, the spirit, in my Nana's voice and with a hand gesture—I remember this so well—reminded me that "men are like buses: if something goes wrong with the forty-three, the forty-four isn't long behind." The forty-three and forty-four were the buses my grandmother used to ride from her house in Hermosa Beach to her business. Nana lived at Hermosa Beach every summer because her guiding goddess was the goddess of the sea, Yemayá, and to honor her, she had sculptures and other sacred representations of her throughout her house.

Only at the time of the adoption of my daughter, who was born a hundred years to the day of my grandmother, did Henriette declare that I did not need her anymore to reach my grandmother's spirit. She did so because she said the significance of the date meant I was now at a different spiritual level, which made it unnecessary to call Nana through her. Sarita was blessed when she was a year old on the day my grandmother died. Only many years later did I find out that Henriette was a deeply respected and well-known *Vodou* priestess.

Obviously, my grandmother believed in spirits. After her husband Warren Kellow died, she spent the rest of her life in regular conversations with his spirit. My grandmother began taking me to these ceremonies to call for her husband's spirit when I was still a little girl. Nana seemed haunted by the tragic death of her stepdaughter, Ruth Gertrude Kellow, and Ruth's mother, Gertrude Kellow.[2] My grandmother consulted a *Candomblé* priestess when she first visited Brazil, where *Candomblé* is a widely practiced religion. She wanted to make sure that their spirits had finally found peace. But it was mainly to my grandfather that she wished to speak. It was with this priestess in Brazil that she first learned how to communicate with spirits. Over the years, she sought out another *Candomblé* priestess to continue this communication. I watched a black Brazilian woman speak in the voice and gestures of a Scots-Irish man—my grandfather. We visited the same woman for ten years. My grandmother made it clear that you could never speak to spirits if you did not respect them and the priestess who brought them forth to you. I respected both. Had I not, I am pretty sure my grandmother would not have let me attend these ceremonies because my presence could only have blocked contact.

My grandmother was always in awe of the process. She certainly revered it. But being the practical woman that she was, Nana's questions to my grandfather, more often than not, were related to daily life and to the running of her business. I remember her asking him whether she should buy a two-color press for the company. The spirit of Warren Kellow was adamant: "Yes, buy the press." Then, in a remarkable Scots-Irish accent, he explained why this investment would put my grandmother "way ahead in the game." Another time, she asked him if she should fire Edith Farrington, the company's accounting vice-president. He did not have an easy answer to that question; there were strengths and weaknesses either way. The problem was that, efficient as Edith Farrington was, she often picked on my grandmother's brother, Gene Coughlin, who worked under her in the accounting department. My grandfather's spirit strongly opposed any abuse of authority. The bottom line was that she should fire her. However, Nana did not go ahead with this and she always felt uneasy about not doing so. She never told me to keep her communications with spirits a secret from my mother, but she did not have to. I knew that she believed that my mother would not understand them. Somehow I also knew that I was there to testify to the wisdom of the spirit of Warren Kellow.[3]

Toward the end of her life, and much to my surprise, my grandmother "came out" about her relationship to my grandfather's ghost to her daughter. She would talk to him in front of my mother. She would ask my mother if she could see her father. My mother thought my grandmother was losing her mind. Yet, in every other way, she seemed sharp as a tack.

My mother had never been as adventurous in her religious practice as my grandmother was, so it was not surprising that my grandmother did not share her devotion to *Candomblé* with my mother. We were raised as lightly Protestant, with the emphasis on *lightly*. But at the time of my grandmother's death, my mother had become a born again Christian. She believed in Jesus. According to my mother, religion should have some proper forms, and my Nana's last rites did not seem to belong to anything my mother could recognize as acceptable Christianity. As her lung disease worsened, my mother turned more and more to Jesus. She prayed every night for a miracle, but the miracle she expected did not come. Despair set in. She blamed herself for not believing enough.

For a time, though, even with waning faith, she continued her version of Christian prayers. She still turned to those prayers at the time when I came back from Paraguay. One week after my return, she had yet another health crisis. She was diagnosed with breast cancer and had to have a mastectomy. The operation was risky because of her lung disease. She was unable to turn to a less drastic surgery like lumpectomy. Radiation, the usual treatment after lumpectomy, was not a possibility for her due to her illness.

My mother and I were not on good terms at the time of her operation. In fact, she was not speaking to me. We had fallen out over my decision to adopt my daughter. She felt strongly that families should look alike, should be made up of people who belonged to the same culture and who had the same color of skin. I had written in my official adoption application documents that I refused to select a child on the basis of a race or color spectrum. Of course, I now understand how ethically complicated an international adoption is.[4] But my mother had her own and different reasons for opposing it. She did not back down. Neither did I. Before I traveled to Paraguay, she made it clear that she wanted nothing to do with this baby and that I was completely unrealistic about what it meant to be a mother.

She turned out to be right. I was unrealistic. Being a mother took so much more sensibility than I could have ever imagined: the vulnerability of mothers in the face of the suffering of their children is enough to overwhelm you. You mourn, again and again, every single failure to protect the child from injuries she might not have needed to experience; you agonize over the constant hardship to make a better world for her. I was particularly unprepared for the change in my perception of the endless small and large injuries that a child of color has to endure in a racist society. I was—and still am—a committed anti-racist. I thought about the intricacies of "white-skin privilege." As I became the mother of my child, I remained on one side of the color line that still very much exists, enough to separate me from my own daughter. Yet sometimes, just sometimes, I could catch a glimmer of how she has to live on the other side of the privileges that I, in spite of myself, still take for granted.

Alicia Ostriker describes the heartrending force of this love. It is so intricately tied to mourning—mourning all that we cannot spare our daughters, even though we know "that whatever doesn't suffer isn't alive":

> When the child begins to suffer, the mother
> Finds in her mouth those burning coals
> You can neither spit out nor swallow. . . .
>
> It is maddening, like the insane yellow
> Of the first blooming forsythia, like a missing
> Limb that goes on hurting the survivor[5]

I was not prepared for "those burning coals." But are any of us ever, really? Was my mother, when she had me, her second child, at the age of twenty-four? My mother obviously did not have to contend with the distress, remorse, and anger of being unable to protect a child of color from vicious racism. But I came to see how, from her point of view, she had to cope with her own sense of powerlessness to protect me from myself. In my refusal to conform to what she understood as basic conventions, my mother saw nothing but a foolish determination to make life difficult for myself. I remember how my unwillingness to respond to bells when I was in the eighth grade led us all the way up to a summons from the superintendent of the schools and, I might add, to consultations with innumerable psychologists who tried to diagnose the "mental illness" that led me to take this particular path of resistance. I remember her yelling at me in frustration as we walked out of the superintendent's office because during my interview with him I had said, "I will not be reduced to one of Pavlov's dogs." I felt misunderstood by someone I thought had allowed herself to be reduced to precisely that: a Pavlovian dog. But once I became a mother, I could begin to imagine how, from her own point of view, making life difficult for myself caused her pain I was unable to understand at the time. I could even see that there could have been a more productive way for me to resist psychological positivism other than by receiving 183 tardy slips and having to spend almost every afternoon of eighth grade in detention.

I wanted my mother to know that I was now open and willing to listen to her tell me how I was often impossible to deal with when I was a child. I was determined to break her silence. I did so by reading the Bible and some of the literature from her Christian group when she was too breathless to read out loud to herself. Slowly, we started to add new prayers and read them aloud. She asked me about the Christian Science prayer I had learned from my grandmother, who used to recite it even long after she turned her heart to *Candomblé*. My mother and I began to say it together. She had started yoga, and we also added a prayer her teacher had taught her:

Take me from the unreal
To the real
From the darkness to the light
From the fear of death to eternal immortality

As we began to talk about our relationship and hers with her own mother, she wanted to know more about Nana's religion. I told her about my experience with the visitations of her father's spirit. We added to our prayers what I now recognize as a version of the ten commandments of Santería.[6] I read her some of my favorite poems and she asked me to include them in what had become our nightly ritual. One of them was by Alicia Ostriker, whose voice is not by coincidence entwined throughout this book. We would end our ritual with a mantra. Whether a poem or a prayer was to become part of the ritual depended on how deeply it affected my mother.

Our prayers changed over time. What had started as my attempt to get my mother to talk to me became a profound spiritual bond between us. We began to practice what we called my "rapping" to her as part of this nightly ritual, and as it became increasingly difficult for her to breathe, my mother relied on me to do the rap while she listened. During the last four years of her life, no matter where I was, I would call her every night at 9:30 Eastern Standard Time, and we would do our "rap" over the phone. She called it "sanctifying reality."

My mother met my daughter when she was fourteen months old and fell in love with my feisty little girl. We went to California two or three times a year during my mother's last years. My mother loved Sarita's artwork. One of

their rituals was that during our visits, my mother would set up an art table
with all sorts of materials so that my daughter could draw pictures to give to
her grandmother. Sarita is an active, boisterous child, but around my mother
she was respectful of her illness and of her need for a quieter kind of compan-
ionship than usually provided. She is a jock and my mother loved that about
her. By the time the two of them met, my mother had already given up all of
her favorite sports, including golf. Nevertheless, she loved to tell Sarita about
her life as an athlete. Sarita, in turn, liked to tell her grandmother about her
own athletic accomplishments.

A year and a half before her grandmother died, my daughter asked to finish our "sanctification of reality" with the prayer I quoted earlier. By her own
choice, Sarita joined our nightly ritual. Given her age, she was deeply and
surprisingly faithful to it. She memorized the prayer and repeated it for my
mother, who loved to hear the sound of her young and strong voice. As my
mother's health worsened, and as she faced the horror of dying of slow suffo-
cation, she began to consider taking her own life. She thought a lot about it,
and ultimately, having read all of the literature that was available to her on
the right to die, she took action in a highly organized way. We relentlessly
debated the issue as it seemed ever more likely that she would make the decision to end her life by her own hand. She consulted her doctors about her future, looking at it with unflinching realism. Finally, one of her doctors agreed
with her that it was her right to decide when the fight was over and that the
need to preserve herself as more than just an ill body was more important
than mere existence. In March 1998, he gave her the pills, which she was to
use in August.

The right to die ceased to be an abstract idea to me. I took every side of
the issue. Not because I was trained as a lawyer, as my mother once accused
me during one of our debates, but because I found myself *emotionally* unable
to take a stand. After all, this was *my mother* who was about to take her life.
The more she argued, the more she set her mind and heart on the idea that
this was what she had to do, the more I admired her. I was the one who faltered in arguments and in conviction. The ferocity and determination that
had once placed us so at odds, now brought us together. When my mother
gleefully told me that she had the pills, I was petrified and started muttering
incoherently. She sternly reminded me of my own work, and using my own
words from *At the Heart of Freedom*, the book I was writing at the time, she
reminded me how important it is for women to demand as a moral right their
"claim to their own person."[7] She argued that, under my formulation, it

made perfect sense[8] to defend the position that extinguishing the actual self was consistent with the kind of moral claim I defended. I came to agree with her that someone could interpret my work that way, but she could not. She did not accept my inconsistency and demanded that I "stand by my guns." My mother had kept my books on her coffee table for years, but I do not believe she ever read them. This manuscript, however, she read in detail. I was pleased. And frightened.

It was my mother who ended the debate. She reached her decision and set the date and the time of her death. She did not want me to sign off passively on her death. She asked me for a very special kind of participation. My mother wanted to die within our "sanctified reality." I was to "rap" to her while she took her pills with a margarita. I was not to break down or the "sanctified reality" would dissolve. In this way, I could accompany her as far as possible on her last journey. She believed that within this "sanctified reality" she would be carried along safely. Loyalty and love demanded that I say yes. This is, philosophically, why I will never be able to debate the right to die. I was called to participate in my mother's death with an intimacy that makes it impossible for me to do anything else but affirm that my mother was "right" in her decision. I invested heart and soul in our "sanctified reality." I did everything I could to keep her protected by believing, as I had done nightly, that there was a spiritual place we could share that bound us together. I rapped and rapped and rapped long after she was no longer consciously with me. And then she was gone.

My mother wore two bracelets on her left wrist. She asked me—as did my grandmother with her rings—to put them on after her death and to never take them off. I did as I was told, "for once," as she joked in her last hours. In her wisdom, my daughter has added the future to my hands. In 1999, she brought me a ring for my right hand, an Irish friendship ring. And two years ago at Christmas, she gave me a bracelet, which she carefully matched to the ones given to me by my mother. I look at the ring and the bracelet and I am always reminded that the next generation is there, "the strong one, the other-after the other."[9]

Just like my grandmother, my mother planned the details of her last days and her funeral. In terms of our own relationship, she also had some things she wanted to make up to me. She regretted her own racism. She was regretful about the time she had failed to spend with Sarita, and was remorseful about what I must have gone through when she withdrew from me at the time when I might have needed her the most—when I was facing the inevitably difficult and stressful process of adoption. Two days before she took her own life, she gave me a check to start The Barbara June Cornell Fund for Inclusion at my daughter's school. She encouraged me to raise money from her friends. We had a laugh together because some of her friends in Orange County, California, were not the kind of people one would imagine contributing to The Barbara June Cornell Fund for Inclusion. My mother wanted the fund to be exclusively for black and Latino/a children. She wanted her friends to know that. Some of her friends were surprised when they received the letter about the fund. They even checked with my brother Brad to make sure that this was really my mother's decision, and not that of her liberal New York City daughter. My brother had to reassure them.

My mother also instructed me not to "yell" at my father when he remarried. She wanted that and firmly stressed that remarriage was best for him. She worried about his health and well-being without a wife. On December 27, 1998, my father married a woman he had met in October, Janet Malcolm, who goes by the name Mimi. I did not yell at him.

The Barbara June Cornell Fund for Inclusion[10] contrasted with my mother's public image. Perhaps her love for me did, too. She thanked me in her last days for bringing out the "woman behind the mask."[11] The fund was to represent her public change of face.

And then there was the memorial service. My father was put in charge of the details of making sure it could be carried out within a reasonable time after her death. She wanted the service to take place at the golf club of which she had long been a member. She chose not to be buried with her parents. Instead, she wished to be cremated and have her ashes scattered over the Pacific Ocean. Her children and grandchildren, and they alone, were to speak at her service. Her grandson Sean, my brother's firstborn son, began the service with the Twenty-third Psalm, and my daughter Sarita, then five years old, ended it with the prayer she had been reciting to my mother every night. All three of her children spoke. As my brother wrote so eloquently in his own eulogy to her, "she thought of us as her greatest legacy." She may not have wanted children in the first place, but once she had us she undoubtedly saw us as her life's work.

My mother died with dignity. I realize that, in choosing the way of her death, she gave me a last gift. By asking me to evoke the "sanctified reality" of our night rituals, she left me with the image of the Barbara June Cornell who had claimed her own person. I was in awe of the woman who died as she did. I saw her full stature and persona shine through the masks that she felt she had worn most of her life.[12] Marie Cardinal, on the other hand, saw her mother as a person only after her death. For me, there is no horrifying death mask to haunt my dreams. For this reason alone, it would have been important to respect her last rites.

There is another easy reason—one recently defended in political theory—to respect a dying person's last wishes. It is that, by doing so, we are respecting his or her "retrospective autonomy."[13] If we violate them in death, we are refusing them the dignity they took with them to their grave.

But there is a more specifically feminist reason for respecting the last wishes of the dead. We—particularly white Anglo Westerners—often contrast the bonds of kinship with the struggle for freedom. Kinship is what ties us down to our past. Freedom lies in choosing our own future. In her beautiful novel *Breath, Eyes, Memory*, Edwidge Danticat provides us with a very different understanding of maternal and feminine genealogy and freedom. Danticat's novel is about the relationship between generations of women. In Haiti, Sophie is raised by her mother's sister Tante Atie until she is twelve years old. Her mother, Martine, regularly sends her tapes about her life in New York. She sends tapes because her sister cannot read. Sophie knows her mother only from those tapes and the bits and pieces of her mother's life that Tante Atie shares with her. That there is a secret in her mother's life, a secret that made her run away from Haiti to New York, swirls around Sophie's memory.

Martine believes she is settled in New York, so much so that she can have her daughter join her. Sophie loves her life with her aunt in Haiti and does not want to leave. But Tante Atie tells her that Martine has always been a wonderful sister and that she will be a great mother. She insists that it is with Martine that Sophie belongs. But there is an awkwardness between mother and daughter long after Sophie arrives in New York. Martine does not have much time to spend with her daughter because she works two jobs. Sophie manages to get only glimpses of her mother's life. She meets her mother's friend, Marc, a Haitian lawyer who helped Martine file the documents that enabled her to bring Sophie to the United States. Once with her, Sophie begins to see how deeply traumatized her mother is, although she does not understand why. Martine has nightmares almost every night. Her daughter watches her as she sleeps, waiting so that she can wake Martine once the nightmares begin:

Whenever my mother was at home, I would stay up all night just waiting for her to have a nightmare. Shortly after she fell asleep, I would hear her screaming for someone to leave her alone. I would run over and shake her as she thrashed about. Her reaction was always the same. When she saw my face, she looked even more frightened.

"*Jesus Marie Joseph.*" She would cover her eyes with her hands. "Sophie, you have saved my life."[14]

The mother will not share her terrible secret with her daughter. She does not want to burden her with the weight of her past. Instead, she has big dreams for Sophie's future, dreams of higher education. Sophie meets her mother's dream and indeed goes to college. Often alone because of her mother's jobs, she meets Joseph, a gentle African American jazz musician who is much older than she and who lives next door. Martine finds out about their relationship and is unable to see it for what it is, fearing that Joseph might take her daughter away. She assumes that he may have taken advantage of her daughter. Sophie swears that nothing has happened, but Martine does not believe her and "tests" her anyway. The test consists of the insertion of her fingers in her daughter's vagina to verify whether she is still a virgin. Sophie had heard about the test from her Tante Atie. Sophie remembers her mother talking as she tried to distract her while she gave her the test:

> "The *Marassas* were two inseparable lovers. They were the same person, duplicated in two. They looked the same, talked the same, walked the same. When they laughed, they even laughed the same and when they cried, their tears were identical. When one went to the stream, the other rushed under the water to get a better look. When one looked in the mirror, the other walked behind the glass to mimic her. What vain lovers they were, those *Marassas*. Admiring one another for being so much alike, for being copies. When you love someone, you want him to be closer to you than your *Marassa*. Closer than your shadow. You want them to be your soul. The more you are alike, the easier this becomes. When you look in a stream, if you saw that man's face, wouldn't you think it was a water spirit? Wouldn't you scream? Wouldn't you think he was hiding under a sheet of water or behind a pane of glass to kill you? The love between mother and daughter is deeper than the sea. You would leave me for an old man who you didn't know the year before. You and I, we could be like *Marassas*. You are giving up a lifetime with me. Do you understand?"
>
> "There are secrets you cannot keep," my mother said after the test.
>
> She pulled a sheet up over my body and walked out of the room with her face buried in her hands. I closed my legs and tried to see Tante Atie's

face. I could understand why she had screamed while her mother had tested her. *There are secrets you cannot keep.*[15]

Sophie resents her mother for putting her through the test. That very night she moves in with Joseph, and shortly thereafter they get married. Her mother cannot forgive them, and silence is all that remains of the dream of a relationship. Sophie has her own baby girl, but the trauma of the test eats her up inside, making her unable to engage in a sexual relationship with her husband. She flees to Haiti to be with her Tante Atie. She returns to Haiti also because her grandmother is getting old, and she wants to make sure she can spend some time with her great granddaughter, Sophie's Brigitte, before she dies. Grandmother and aunt insist that she must try to find a way back to her husband and especially to her mother. Sophie reluctantly returns to New York, and slowly, very slowly, she and her mother struggle together to reconnect. One day, much to Sophie's surprise, her mother announces that she is pregnant. She shares with her daughter her doubts about having the baby even though Marc, her longtime friend, is willing to support her and the baby. During her pregnancy, the nightmares increase. She describes to her daughter how she is struggling to stay one step ahead of mental institutionalization.

Finally, Martine succumbs to the nightmare world that haunts her pregnancy and takes her own life by stabbing herself repeatedly in the stomach over and over again. She wants no more babies born into bitterness and secrets. Marc finds her in the bathroom after she has stabbed herself seventeen times. Still conscious as she is taken to the hospital, she keeps repeating, over and over again, that she simply could not have this baby.

Martine is taken back to Haiti for burial, and Marc and Sophie go with her. Joseph stays home to take care of Brigitte. As Martine is laid in the ground, her secret is finally revealed: she had been brutally raped in the sugar-cane fields near her home at the age of sixteen by a *Tonton Macoute*.[16] Sophie was the result of that rape. Sophie chooses a crimson dress for her mother's burial gown, a ritual reminder to herself and the other mourners of the woman's blood that had been spilled.

After her mother's funeral, Sophie runs madly into the sugar-cane fields looking for the spot where her mother's dignity had been so horribly violated. It had become "a place where nightmares are passed through the generations like heirlooms."[17] Sophie will later remember: "I come from a place where breath, eyes, and memory are one, a place from which you carry your past like the hair on your head. Where women return to their children as butterflies or as tears in the eyes of the statues that their daughters pray to."[18] But the girl at the scene of the trauma cannot speak. Her grandmother approaches her gently. In Creole, "*Ou libéré?*" is what the women shout to each other when they are carrying their goods on the head. It means, "Are you managing your burden? If not, then another woman will help you." It also means, "Are you free?" As Sophie stands there, choking on her sorrow, her grandmother speaks:

> "Listen. Listen before it passes, *Paròl gin pié zèl*. The words can give wings to your feet. There is so much to say, but time has failed you," she said. "There is a place where women are buried in clothes the color of flames, where we drop coffee on the ground for those who went ahead, where the daughter is never fully a woman until her mother has passed on before her. There is always a place where, if you listen closely in the night, you will hear your mother telling a story and at the end of the tale, she will ask you this question: '*Ou libéré?*' Are you free, my daughter?"
>
> My grandmother quickly pressed her fingers over my lips.
>
> "Now," she said, "you will know how to answer."[19]

After my mother's death I read the novel *Breath, Eyes, Memory* over and over again. Some of the beautiful Creole phrases scattered throughout the book took me back to rituals with Henriette, from whom I had first heard them. The week after my mother died, I sought solace from Henriette in a ritual especially dedicated to one's dead mother. The question, "*Ou libéré?*" haunted me. I wanted Henriette simply to answer the question about my mother. Was my mother free? "*Ou libéré? Ou libéré?*" Like a chant, the question kept coming back to me. For my mother and for myself, I asked Henriette to give me a simple answer. She smiled and said nothing. I kept writing the book.

I have always known this book was the offering that my mother wanted me to give to her spirit. And so, Barbara June Kellow Cornell, I now offer you your book. I now *do* know how to answer the question, "*Ou libéré?*," for both my mother and for myself: It was you, Mother, who demanded the book. It was you who insisted on the words, "*Paròl gin pié zèl*." The words you asked for have given wings to my feet. I am taking up my burden, my place in my maternal line, and I have claimed my freedom. Because you insisted I muster the courage to write this book, you have brought me to a new threshold. Thank you. May you rest in peace.

Appendices

INTERVIEWS / SPANISH VERSION

Appendix A

ZONIA VILLANUEVA
MAYO 23, 2000
(ENTREVISTA EN ESPAÑOL)

Drucilla Cornell: [Zonia], ¿De dónde es usted originalmente?

Zonia Villanueva: Soy de El Salvador del departamento de Santa Ana.

D.C.: ¿Cómo se llama usted?

Z.V.: Zonia Villanueva. Villanueva de casada y Valle de soltera.

D.C.: Zonia, ¿por qué vino usted a los Estados Unidos y cuándo?

Z.V.: Yo vine de El Salvador en el ochenta y ocho [1988] a los Estados Unidos porque la situación [de guerra] para nosotros era muy difícil, para mantenernos. Porque no teníamos un trabajo que nos diera la cantidad de dinero para nosotros poder vivir. Entonces, yo decidí venirme. Tenía mis hermanos acá . . . uno . . . y le conté que me quería venir, y el me dijo que me viniera. Entonces hice el viaje en el ochenta y ocho, y me vine a los Estados Unidos.

D.C.: ¿Es su hermano ciudadano de los Estados Unidos?

Z.V.: Sí, mi hermano es ciudadano de acá de los Estados Unidos.

D.C.: Zonia, ¿es [usted] ahora ciudadana de los Estados Unidos?

Constanza Morales-Mair: ¿Ya logró sacar documentos de ciudadanía?

Z.V.: Yo estoy en proceso. Tengo el permiso de trabajo como por siete años. Y ahorita estoy en proceso para la residencia, a ver que pasa después.

C.M.M.: Zonia, usted está aplicando para la residencia, verdad?

Z.V.: Sí, así es.

C.M.M. [*a D.C.*]: Ha presentado papeles y está solicitando la residencia, para la "green card"

D.C.: Su hermano presentó papeles por usted?

Z.V.: No, no. Tengo el permiso de trabajo por el PTC que ofreció antes [la oficina de] Inmigración a los refugiados. Un programa que ellos sacaron—el PTC—para comenzar uno a tener permiso para trabajar.

C.M.M.: Zonia, ¿qué quiere decir PTC?

Z.V.: No sé, un documento que nos dieron a hacer para sacar el primer permiso y que nos lo dieron a todos los salvadoreños.

C.M.M.: ¿Para todos los salvadoreños?

Z.V.: A todos los salvadoreños [refugiados].

D.C.: ¿Ha trabajado usted como "nanny" [como niñera] en los Estados Unidos?

Z.V.: Sí. Trabajé en el ochenta y ocho [1988]. Los primeros meses que yo vine de El Salvador, yo trabajé cuidando tres niños. Trabajé con una familia salvadoreña, entonces... muy cierto, ellos me pagaban bien poco pero me hicieron sentirme bien, pues pasé bien el tiempo que yo estuve con ellos, [que fue] como unos ocho meses. Yo trabajé cuidando niños: una niña de cinco meses, una niña como de año y medio, y una niña como de tres años.

D.C.: ¿Por qué se fue de ese trabajo?

Z.V.: Yo no me fui del trabajo, sino que la misma señora me consiguió... este... Ella me quería mucho entonces ella quería que yo ganara más. Ella me buscó trabajo en una fábrica de lápices de labios. Entonces para yo desaburrirme y me gustaba, entonces yo... ella me dice que en la tarde yo le cuide las niñas y que fuera por el día a la fábrica, entonces [ese] fue mi primer trabajo en una fábrica; una fábrica de lápices labiales.

D.C.: ¿Por cuánto tiempo trabajó en dos empleos?

Z.V.: Yo vivía ahí mismo en la casa de ella. Cuando salía yo me quedaba con las niñas, ya era algo... una confianza que teníamos, ¿no? Entonces trabajé como... como un año yo trabajé así con ellas. Y, después, vino mi esposo de El Salvador. Ya yo dejé de trabajar así y después yo me mudé para acá porque allá es Suffolk County. Y aquí es Nassau County, entonces me mudé de Suffolk para Nassau. Entonces ya después yo seguí trabajando en una fábrica. Trabajé en una fábrica de plásticos, después, empacando bolsas en otra fábrica. Después, este, yo le hice la pregunta a una señora, [le dije] que me gustaba cuidar ancianos. Entonces ella me dijo que eso era fácil, que fuera a hacer un "training."

C.M.M.: Zonia, ¿fue muy pesado trabajar en dos sitios? ¿Le pareció mal el arreglo?

Z.V.: No, no. Yo pienso que fue algo que ella quería hacer para que yo me ambientara, porque como estaba recién venida, era que ella quería que yo me ambientara al clima y todo, salir afuera y agarrara aire. Yo no lo tomo como si fuera algo costoso. Fue costoso para mí porque yo no manejo, andaba en bus publico, ¿no? Y ella, cuando podía, pues, me iba a recoger.

D.C.: ¿Su hermana vino con su esposo?

Z.V.: ¿Mi hermana? No. Mi hermana... Sí vino una hermana. Ella tenía el esposo acá también, pero estaba otro hermano que había venido también, así... por la frontera, ¿no? y otro hermano. Total, habíamos tres hermanos: tres varones y cuatro hembras acá. Y ahora, pues, mi hermano mayor es ciudadano. Y ellos, él y su esposa, pidieron a mis padres. Les salió la residencia y, entonces, ellos están acá.

D.C.: ¿Tu hermana está contigo siempre?

Z.V.: No juntas en la misma casa, pero sí estamos en buena comunicación. O sea que tienen su apartamento cada una, ¿verdad? Pero, sí, cuando hay alguna celebración nos juntamos y nos sentimos bien en familia . . . ahora que ha venido Mamá y todos estamos unidos, como Dios quiere.

C.M.M. [*a D.C.*]: Lo que preguntó fue que . . .

D.C. [*a C.M.M.*]: ¿Estaba su hija siempre con ella?

C.M.M.: No. En realidad, lo que usted preguntó fue si su hermana estaba siempre con ella.

D.C.: Oh! ¡Whoops! (*risa*) Whoops! ¡Tu hija, tu hija!

Z.V.: Mi hija es nacida acá. Ella nació en 1990. Ahora tiene ya unos diez años y me dice que no tengo que preocuparme porque ella es ciudadana norteamericana y no va a permitir que nos manden a El Salvador.

D.C.: ¿Tu hija nació aquí después de que llegó tu marido?

Z.V.: Sí. Yo vine a principios del ochenta y ocho, y mi esposo vino a finales del ochenta y ocho.

C.M.M.: Casi un año después.

Z.V.: Sí. Un año después.

C.M.M. Y la niña nació casi dos años después de que él vino.

Z.V.: Sí, dos años después.

D.C.: ¡Una bella hija!

C.M.M.: ¿Cómo se llama?

Z.V.: Jennifer.

D.C.: ¿Cuántos años [tiene]?

Z.V.: Diez

D.C.: Mi hija tiene siete años. ¿Por qué vino Zonia al al *Workplace Project*?

Z.V.: Bueno, pero y . . . ¿Cuento la historia de mi trabajo con los ancianos?

C.M.M. [*a D.C.*]: Zonia había mencionado el hecho de que había trabajado con los ancianos y no hemos vuelto al tema. ¿Quieres saber cómo consiguió ese trabajo?

D.C.: ¡Oh! ¡Sí!

C.M.M.: Bueno, Zonia, entonces por favor cuéntenos de su trabajo con los ancianos y, luego, cómo llegó a vincularse con el Centro de Trabajo.

Z.V.: Yo . . . este . . . mi ilusión era tener un trabajo mejor y ganar un poquito más. Yo siempre tengo algo [dentro de mí] que digo: "Yo puedo, yo puedo," aunque no pueda, pero esa es mi opción de decir: "yo puedo hacerlo." Entonces yo quise sacar el "training" de cuidar ancianos, aunque no sabía mucho inglés, pero me fui a registrar en una agencia y saqué el certificado y todo. Entonces ya trabajé cuidando ancianos desde el noventa y dos hasta el noventa y ocho. Cuidando ancianos . . . Me cambiaba de agencia porque, lo mismo, yo quería [ir] adonde me pagaran un poquito más. Entonces, yo cuidé ancianos como por seis años. Y . . . bueno . . . hasta que hubo un cambio en el cuidado de los ancianos por la razón de que el horario . . . los ancianitos se quedaron sin seguros, sin Medicare que rebajó tantas horas, entonces para mí era bien difícil ya tener un caso de ocho horas o diez horas diarias. Ya eran tres, cuatro horas. Para hacer ocho horas tenía que ver como cinco pacientes. Trabajar con cinco pacientes . . . ir a trabajar con cinco pacientes para yo tener un trabajo completo. Entonces, de esa manera, yo dejé la escuela de ancianos y en el noventa y ocho, en la Iglesia de [la Virgen de] Loretto tiraron unos papeles diciendo que quien trabajara o quería trabajar en limpieza de casas que se pusiera en contacto, ¿verdad? Entonces mi hermana agarró ese papel y yo no lo agarré y me dice mi hermana: "Fíjate que me llamaron, si querés vamos," me dice y yo le digo: "¡Pero yo no llené el papel!" . . ."¡Ah, vamos!" [dijo ella]. Y entonces, como yo siempre soy bien "meque," fui con ella. Entonces ya vino un señor de Long Island ya diciendo . . .

C.M.M.: Disculpe, Zonia, paremos aquí un momento . . . Drucilla, ¿está usted entendiendo?

D.C.: Mucho, creo. Zonia, sigue. Muy bien, muy bien.

Z.V.: Entonces es así como fue que yo me quedé ahí y entonces el señor de Long Island dijo que él quería que . . . como que hiciéramos algo entre ellos y la comunidad de Hempstead, de Loretto, que hiciéramos algo como de un grupo, como de limpieza de casas, como un "day-care" de niños. Algo para poder tener trabajo. No era que había el trabajo ¿verdad? Entonces cuando, en el noventa y ocho, Nadia Marín, que es la ejecutiva directora del Centro de Derechos Laborales estaba de "translations"— traduciendo—entonces, bueno, habemos muchas que vamos . . . [y] que creemos que ya el día de mañana hay trabajo. Entonces, ese día fue bien duro porque la gente pensó que ella iba a decir: "Tú vas a ir mañana a trabajar," y no fue así, sino que fue a través de un proyecto, ¿verdad? Entonces, recuerdo que ahí ya se anotaron los nombres de las que quedamos en la reunión, que eramos como cinco nada más, que quedamos en esa primera reunión. Entonces dijo Nadia que aquí en el Centro de Derechos Laborales también estaban tratando de ayudar a las mujeres, ¿verdad? Para conseguir un buen trabajo. Entonces, bueno, eso pasó. Como a los tres, cuatro meses yo recibí una llamada que me decía que me llamaban del Centro de Derechos Laborales. Entonces fue como yo vine al Centro de Derechos Laborales y ya me dijeron que había sido una persona de las que había estado en Loretto, escuchando, y que yo sí quería ayudar y que me ayudaran a tener un buen trabajo. Entonces así fue como ya formamos una *Cooperativa Unidad*, que se formaron los cuatro talleres para tener un poquito más de respeto y dignidad en el trabajo, porque eso es lo que nosotros queríamos ¿no? Tener, este . . . como un trabajo más fijo y un sueldo mejor. Por eso yo llegué aquí al Centro de Derechos Laborales, y gracias a Dios el grupo que hemos formado—vamos para dos años y medio—y hemos logrado tener, pues por lo menos cinco casas cada una. Yo me siento que es un trabajo digno porque yo soy la . . . el horario que yo me pongo, no, sí son tres horas y cuatro horas. Son mis sesenta dólares y me voy para la casa. Y tengo tiempo para recoger a mi hija. Tengo más tiempo para hacer algo más que yo quiera dentro de la casa. En cambio, trabajar en una fábrica: trabajo ocho horas a diario por cuarenta y cinco dólares al día. Son ocho horas a diario y talvez parada. Fui a trabajar a una fábrica—es algo bien duro que me pasó entre el noventa y siete al noventa y ocho—fui a trabajar a una fábrica, y me pusieron a sacar unas sierritas calientes—de hierro—todo eso caliente de unas máquinas, con guantes y todo y entonces yo sólo me mantenía herida. Así. Entonces—yo soy diabética—entonces yo le decía al señor que yo no podía trabajar porque ya estaba muy herida de los dedos, y aquello sucio y todo, y le dije que era enferma, que tenía diabetes y me daba miedo por las heridas y, entonces, que si no había otra posición de otra cosa. Y me dijo él . . . Él quiso ayudarme porque me dijo que había tratado de decirle a otro "manager" para que me pasaran a empacar, y entonces me dijo que el "manager" le había dicho que no, que no había . . . Bueno, entonces sólo trabajé dos . . . dos . . . tres días trabajé en esa fábrica porque era de material fuerte, de hierro, entonces . . . y ahora, pues ya tengo dos años de estar trabajando con la *Cooperativa Unity* que hemos formado, porque, aunque no tengamos voto, pero tenemos como un poquito más de respeto. No nos hincan, porque hay muchas partes de trabajo que lo hincan a uno a limpiar pisos, a limpiar, este . . . cosas de los perros y todo eso entonces, por lo menos, eso no . . . es el respeto de nosotros que sentimos, porque no lo hacemos. Nos han enseñado a hacernos valer como mujeres, trabajadoras que somos con dignidad y seguridad.

D.C.: Yo también he trabajado en fábricas y he visto muchos accidentes. A una mujer que trabajaba junto a mí, una máquina le cortó la mano.

Z.V.: ¡Ay, Dios mío!

D.C.: Hubo sangre por todas partes. Corrimos a llevarla al hospital pero cuando llegamos allí y los médicos nos pidieron la mano. Así es que otras compañeras y yo fuimos a la fábrica a buscar la mano. Estaba triturada. El jefe la había tirado al basurero. Así es que otra trabajadora me sostuvo por los pies sobre la basura, y busqué la mano . . .

Z.V.: ¡Ay, Jesús!

D.C.: . . . y encontré la mano, pero estaba tan triturada que no hubo nada que hacer.

Z.V.: Esas son muchas historias que se ven ahora que mucha gente pasa en las fábricas. Así son muchas de las historias de los inmigrantes que sufren aquí, ya sean mujer u hombre, que trabajan en fábricas tienen muchos accidentes, mucho peligro porque en muchas fábricas hay mucha seguridad, de los químicos y eso, pero en otras . . . no hay nada . . . sí.

[Durante la traducción le preguntamos a Zonia si las mujeres que aparecen en las fotografías expuestas en las paredes de la oficina del Centro son las mismas mujeres que se presentaron y se quedaron en la primera reunión. Nos hace notar que esas fotos no son sólo de la mujeres que estuvieron ahí, sino que ahora son muchas más.]

Z.V.: Esa foto ya fue cuando la otra reunión que hicimos acá en la que ya habíamos más ¿verdad? Pero a través de las reuniones, la gente hacía lo mismo. Vienen cincuenta, y se quedan diez o quince.

D.C.: ¿Hace cuántos años fue esto?

Z.V.: Hace dos años. Desde el noventa y ocho hasta esta fecha.

D.C.: Zonia, ¿ahora trabaja solamente a través de este programa?

Z.V.: Sí. Solamente.

D.C.: Zonia, dos preguntas: Uno: ¿cómo consiguen los trabajos? Y dos: ¿cuál es su programa de reforma? ¿Cuál es su posición frente a la amnistía, por ejemplo?

Z.V.: Pues apoyando. O sea, como ya estamos dentro del Centro de Derechos Laborales, y si hay una marcha para amnistía, entonces nos invitan y podemos ir, porque la necesitamos también, ¿verdad? Para una amnistía, o si hay que ir a apoyar . . . la vez pasada fuimos a Washington para pedir la residencia. Fuimos a marchar. Yo he estado apoyando porque me gusta y siento que si yo lo puedo hacer, algunos otros lo podrán hacer por otros que vengan. Y entonces . . . y [en cuanto a] la distribución de trabajo que se hace aquí en la *Unity*: primeramente, formamos cuatro talleres. Ahí nos dieron a conocer cuáles son . . . este . . . ¿por qué nos organizamos? ¿qué es lo que logramos organizadas? ¿qué es lo que . . . al unirse un grupo . . . ¿cuál es el resultado? Aprendimos a conocernos, primeramente, un grupo de mujeres acá en el Centro de Derechos Laborales. Aprendimos a conocernos en esos cuatro talleres, aprendimos cuáles son los derechos, y entonces aprendimos a saber el respeto que nos merecemos como mujeres trabajadoras. Y para distribuir el trabajo a la compañeras, todo está basado en reglas ¿no? Porque no todos somos responsables. Queremos trabajo pero nosotras tenemos que ser responsables. Esto lo hicimos por un reglamento de asistencia: la que cumplió los cuatro talleres, fue la que fue a limpiar la primerita casa de la *Unity*.

C.M.M. [*a D.C.*]: La regla es que si no terminas los cuatro talleres, no se te asigna trabajo porque de lo que se trata, también, es de aprender a conocer la organización y . . .

[*A Zonia*]: . . . y también son talleres de enseñanza de la limpieza ¿cierto? Eso también está incluído en los talleres?

Z.V.: Sí, también. Los cuatro primeros talleres que hicimos fueron para conocernos, aprender y desarrollarnos, para [aprender] cómo hablar con el cliente, para aprender uno cómo se organiza un grupo de mujeres, qué hace un grupo de mujeres organizadas . . . Esos fueron los primeros cuatro talleres. Después de esos cuatro talleres, formamos cuatro comités ¿verdad? Entonces es un reglamento que se dio bien bonito, porque aprendimos mucho y después nos enseñaron que de esos cuatro talleres salieron cuatro comités que son: Reglas, Publicidad, Educación, y Finanzas. Entonces cada comité se encargó de su trabajo. Por ejemplo, el Comité de Reglas: poner el reglamento de asistencia [y el orden de llegada de casas, a quién se le iba a dar la primera casa [y] por qué se le iba a dar a esa primera casa . . . Un reglamento ¿verdad? Y Publicidad se encargó de hacer la publicidad: sacar los "flyers," ir a repartirlos, hacer publicidad para que vinieran las casas. Educación se encargó de hacer los talleres teóricos, taller de limpieza en una casa, a enseñarnos cuáles son los líquidos que se usan, qué es lo que se tenía que hacer dentro de la casa y cómo el cliente lo podía recibir. Tenemos una señora voluntaria, que, a lo mejor ustedes la conocieron, que es Nancy Ryan. Ella nos presta su casa para el entrenamiento.

D.C.: ¡Por eso es que su casa siempre está limpia!

C.M.M.: Si, ¿se acuerda que ella dijo que su casa siempre estaba limpia?

D.C.: Así es que, ¿para conseguir un trabajo a través de la cooperativa hay que ser de *Unidad*?

Z.V.: Sí. Hay que ser miembro de la *Cooperativa Unidad* y cumplir los cuatro talleres—que yo pasé esos cuatro talleres—y pertenecer a un comité. Al que le guste. Para, así mismo, ver la asistencia que tiene, el orden de llegada, si es puntual, así mismo se le va dando [trabajo].

D.C.: ¿Cuántas mujeres hay en *Unity* ahora?

Z.V.: Aproximadamente, como veinticinco.

D.C.: ¡Muy bien!

Z.V.: Este cuadro es la séptima ronda. Ahí se ve por orden de asistencia, por orden de asistencia de los comités [*Zonia señala el tablero donde están especificados los turnos de las personas y las casas que les son asignadas*] Entonces cuando se dio la primera casa, el 3 de febrero del noventa y nueve, esa casa me tocó a mí. Era la primerita.

C.M.M.: Usted fue la que completó los cuatro talleres.

Z.V.: Sí, yo fui la que completó los cuatro talleres, y siempre venía y venía y venía y por eso me eligieron la primera.

C.M.M.: ¿Cuántas casas tienen registradas?

Z.V.: Aproximadamente hay como unas setenta casas. Eso es variable porque en cuanto vienen casas, llaman que quieren alguien y a las dos, tres semanas cancelan. Sí. Algunas entran y otras salen.

D.C.: ¿Ustedes hacen sólo limpieza de casas o cuidan niños?

Z.V.: Cleaning. No hacemos "child care." Tenemos, por ejemplo, una compañera que la pusieron a limpiar el piso de la cocina que es bien grande, y querían que se hincara con un cepillo, sin "mop" y sin nada. Así con el cepillo en el piso, arrastrándose ¿no? Entonces lo hizo la primera vez y le pidió un "mop" a la señora. Pues no le dio el "mop" ni nada, y le dijo que así ya no le trabajaba. Entonces se llamó y se canceló porque lo que queremos es que no hinquen a las mujeres. Siempre uno tiene . . . pues experiencias buenas y experiencias malas ¿no? Con cada cosa uno aprende.

D.C.: Zonia. ¿Son ustedes feministas? ¿Se consideran ustedes feministas?

Z.V.: Sí.

D.C.: ¿Hacia dónde le gustaría ver llegar el feminismo? ¿Cuál o cómo es su feminismo?

Z.V.: ¿Como mujer? Me gustaría tener un triunfo, ¿no? Como mujer que soy y como feminista que fuera, me ha gustado tratar, de trabajos en trabajos, a ver adonde yo puedo salir mejor adelante. Una intención que tuve también, porque me encanta muchísimo ser cosmetóloga, hacer peinados y eso, y lo intenté aquí pero no pude porque no tenía mi residencia. Y sí, me llama la atención un día poder estudiar y sacarla [la licencia]. Sería lo primero que haría porque me llama muchísimo . . . [es] que quiero hacerlo. Y no pierdo la esperanza en que al salir los papeles lo pueda hacer.

D.C.: Zonia, una última pregunta. ¿Qué sueños tiene usted para su hija?

Z.V.: Los sueños míos para mi hija . . . Que estudie lo que más pueda y que llegue a ser algo en el futuro, como una abogada o una doctora, que eso es lo que ella dice que quiere ser: una doctora latina . . . una pediatra, dice. Mis sueños son esos: poderle dar el estudio para lo que esté a mi alcance. Para que ella pueda llegar a hacer algo en el futuro. Y no como yo, que quede limpiando casas o haciendo trabajitos. Hace falta el estudio porque uno viene a estos países de acá . . . Aquí hay muchas oportunidades pero la situación de llegada de uno es bien difícil porque uno viene con aquella cosa de que quiere trabajar, que quiere hacer la plata para mandarle a los familiares.

D.C.: Muchas gracias, Zonia. Ahora . . . ¿Vamos a comer?

Z.V.: Sí, pero antes, también tengo una historia de cuando cuidaba ancianos. Me mandaron a una casa de una persona morena. Y entonces cuando yo fui y ella, cuando me vio, me dijo: "¡Oh, no, tú eres hispana!"—dice—y no me dejó entrar. Entonces, ahí sí me mató porque yo me puse a llorar. Y me quedé yo así, con aquel gran frío afuera ¿no? Me quedé con un pie adentro y otro afuera y dice: "¡No, tú eres hispana, 'you don't understand!'" Me dijo, que yo no entendía. Y ella no me había hablado absolutamente nada a mí. Ella no sabía si yo le entendía o qué. Y yo le dije: "Please let me in!" Le dije yo que quería ir adentro, para usar el teléfono y avisar a la agencia y decirles que ella me estaba rechazando, pero a todo esto, yo estaba llorando porque me había sentido discriminada, o no sé, pero sentía yo . . . algo fuerte que había pasado y yo me puse a llorar y ella me vio, y ella no me abrió la puerta. Sólo la dejó así [entreabierta]. Entonces yo me vine para la agencia y les dije que cuando me mandaran a un caso, que preguntaran que tenían a alguien disponible pero que era latino, si la querían, sí o no. Porque . . . uno se siente bien mal. ¡Discriminado! Y le digo yo: "Usted no me ha hablado para ver si yo le entiendo."

D.C.: Afro-americanos son discriminados y a su vez también discriminan. Yo he trabajado como organizadora de sindicatos y, a veces, había tensión entre latinos y afroamericanos. Una situación muy difícil. Nosotras, las mujeres del comité organizador, trabajamos muy duro por la solidaridad del movimiento pero ésta era muy frágil.

Appendix B

Mi nombre es Antonieta. Soy de El Salvador. Emigré a este país en Junio de 1988; acompañada de mis hijos de 16 y 7 años respectivamente.

Esta decision la tomé a raíz del problema sociopolítico en El Salvador, pues vivir en una guerra no es fácil. Soy de una ciudad del interior del país, para ser más precisos del oriente del país, donde el conflicto armado fue más fuerte y los enfrentamientos entre guerrilla y ejército ponian en riesgo la vida de las personas que ahí vivíamos y además la guerrilla reclutaba a los jóvenes; para proteger a mis hijos fue la razón por la que emigré. Seguimos en este país porque aunque ya la guerra terminó, allá quedaron problemas de delincuencia de todo tipo, no hay respeto a la vida y aquí nos sentimos más seguros y, con doce años viviendo aquí, ya no queremos regresar. Ya mis hijos están acostumbrados a este país. Esto no quiere decir que hayamos olvidado nuestra tierra, pues aunque la mayoría de nuestra gente es trabajadora, respetuosa y quiere lo mejor para el pueblo, a veces las cosas se dan de otra manera y la falta de trabajo, la superpoblación; y jóvenes que quedaron huérfanos, niños que crecieron en medio de la guerra, esto los ha llevado a resolver su vida con violencia. Dios quiera que un día nuestro país recupere la paz.

Le voy a hablar un poco de mí. En El Salvador yo trabajaba dando clases en un bachillerato y como gerente en una cooperativa de caficultores y viajé para esta nación con una visa de turista. Aquí se encuentran algunos parientes o familiares y con ellos disfruté conociendo lo bello que es Nueva York. Pero como no podía pasar sólo gastando los dólares que traía, producto de mis ahorros, decidí buscar trabajo como doméstica, ya que en otro tipo de empleo necesitaba tener permiso para trabajar. Bien. Comencé a buscar en los periódicos y leí un anuncio que decía "Se necesita señora para dormir adentro, cuidar niños y limpiar, lavar, planchar, cocinar, etc. No necesita inglés." Yo dije:

"Esto es lo ideal para mi." Hablé por teléfono, me dio una cita, me entrevistó, revisó mis documentos de identidad y me dió el trabajo de inmediato. A lo que yo le dije que su casa estaría limpia y, lo que para mí era lo más importante, que sus niños estarían seguros física y emocionalmente porque los iba a cuidar como míos. Lamentablemente la señora no sabía como guiar a la niña. A mí me decía: "No la dejes que ande de casa en casa en el vecindario," y cuando yo le decía a la niña que en ese momento no podía salir porque su hermanito estaba dormido o comiendo, y yo no podía acompañarla, ella le hablaba por teléfono a su mamá quejándose porque yo no la dejaba salir sola a casa de sus amigos. Y era cuando la mamá le decía: "Sí, vé tú sola." Y era un conflícto pues yo quedaba ante la pequeña de siete años como alguien no grato para ella y me decía palabras ofensivas, me hacía señas con los dedos, etc. Esto me hizo renunciar al trabajo pues temía le pasara algo en la calle y luego yo podía ser responsabilizada. Lo sentí ya que me había encariñado con el bebé de dos años de edad. Le hice ver a la señora el por qué de mi renuncia y que esperaba ella supiera en el futuro cuándo decirle "no" a su hijita. Hablo de todo esto porque, si alguien lee estas líneas, que sepa que pueden confiar o apoyarse en las personas que les damos un servicio y que a los hijos hay que amarlos y protegerlos, pero no decirles "sí" a todo lo que ellos quieran pues, cuando de adolescentes se les quiere corregir, creo se dan los conflictos y el resultado es la violencia en la calle y las escuelas. No hay que usar el maltrato físico o emocional para guiar a los hijos. Bien.

Después de dejar este trabajo, una conocida me dice: "Hay una señora que necesita una persona para trabajar interna y encargarse de todo en la casa. Paga ciento sesenta dólares por cinco días y puede darte los papeles" [patrocinar para legalizar residencia en este país] y me presenté a la entrevista. Al día siguiente comencé mi nuevo trabajo [Sept. 2 de 1988]. La familia era de cuatro personas: los padres, un niño de 12 años y una niña de 11 años. Muy buenas personas en el trato, el trabajo era cocinar, y mantener la casa limpia, ordenada, lavar y planchar. Yo cumplía con todas mis obligaciones. Cuando ya tenía un par de meses con ellos, comencé a tramitar mis papeles con un abogado recomendado por la señora y esto me llevó a un gasto de mil doscientos dólares, con la esperanza de legalizarme, y lo hacía más que todo por el futuro de mis hijos. Pero lo triste fue que a finales de 1994 [seis años después] me dijo la señora que no podía darme la firma pues no quería tener problemas por tener una indocumentada. Aquí me sentí abusada, utilizada, engañada, ya que me había privado de estar cerca de mis hijos. Si acepté un salario de ciento sesenta dólares fue porque me sentía agradecida porque iban a ayudarme a que un día me legalizara. Económicamente no estaba bien pues tenía que pagar renta, comida, etc. y, lo que es peor: mis hijos solos en el apartamento. Sólo los fines de semana pasaba yo con ellos. Fue un sacrificio en vano y gracias a Dios que me dió unos buenos hijos y ellos correspondieron portándose correctamente, ya que no se divagaron de la casa a la escuela y viceversa, y el mayor trabajaba y estudiaba y cuidaba a su hermanito. Puede imaginarse las angustias que pasaba pensando en los peligros que hay alrededor, sólo orando para que nada les pasara y que escucharan ellos (mis hijos) mis consejos, ya que, como soy divorciada, no tenían un padre cerca. Imagínese lo que pudiera haber pasado a dos menores solos. Y perdí la oportunidad y el tiempo, pues mi hermano me hubiese pedido.

Usted me pregunta si recibí prestaciones sociales. No. Nunca tuve ninguna clase de beneficios . . . Drucilla, me pregunta usted si yo alguna vez les pregunté a las personas con las que trabajaba si podía tener a mis hijos conmigo. Ciertamente nunca les pregunté, pero sí, de antemano, sabía que no era possible eso. Uno a veces sabe qué esperar de la gente. Déjeme darle una idea: el señor es dentista y una vez necesité de sus servicios ya que tenía una pieza dental dañada y él me atendió y me cobró por sus servicios. Cierto día les pedí aumento y me dijeron que no podían ya que me daban casa, y les expliqué que eso para mí no era un ahorro ya que yo pagaba un apartamento para que vivieran mis hijos. Y aún les pedí que me dejaran salir todas las noches con la promesa de que yo estaría todos los días temprano y se negaron diciéndome que para darme los papeles como "housekeeper" tenía que dormir allí.

D.C.: A qué hora tendrías que entrar a trabajar?

Antonieta: A las 7 A.M.

D.C.: A qué hora podrías haber salido?

Antonieta: A las 8 ó 9 P.M., después de servir la cena y limpiar la cocina.

Como verán, no es fácil para los inmigrantes salir adelante.

¿Cómo llegué a conocer el Centro de Derechos Laborales? Cierto día leí un anuncio en el periódico que el Centro ofrecía un curso de Promotores Laborales, y que consistía en siete clases de dos horas de duración cada clase, para conocer sobre las leyes laborales y organización. Me interesó pues tendría conocimientos sobre mis derechos y podría llevarle el mensaje a otra persona, pues nos ayuda a evitar el abuso y a que nosotros cumplamos con nuestras responsabilidades. Después de recibir mi diploma—esto fue en 1993—no pude involucrarme mucho en el trabajo del Centro porque aún trabajaba interna pero, después, cuando ya no estaba interna, me volví voluntaria activa. He sido miembro de la Junta Directiva del Centro.

D.C.: Estoy tan conmovida . . . El mundo tiene que cambiar para que esto no siga sucediendo.

Antonieta: Yo creo que para que esto suceda las personas que necesitan de nuestros servicios, deben humanizarse y tomar conciencia de que, como seres humanos, tenemos derechos.

Constanza Morales-Mair: Antonieta, ¿cómo es que su hermano obtuvo la ciudadanía y pudo solicitarle status de residencia?

Antonieta: Mi hermano llegó a este país hace mas de treinta años y mí mamá hace unos veinte años o más. Se hicieron residentes por medio de un amnistía y actualmente mi hermano es ciudadano americano.

C.M.M.: Es decir que, a través de su hermano y su mamá, habría sido más fácil conseguir sus papeles de residencia.

Antonieta: ¡A esta fecha y me habrían salido! Ya que después de esos años perdidos tuve que que aplicar de nuevo. Mi mamá me está patrocinando; y si no lo hicimos así desde el principio fue porque me aconsejaron que por medio de mis patrones sería más rapido, por ser ellos ciudadanos y personas con dinero que necesitaban mis servicios; que en esa categoría había más visas disponibles.

C.M.M.: Nos decía que el trabajo en el centro y transmitir el mensaje a otras mujeres es una forma de cambiar el mundo y que sentir eso la motivó a continuar . . .

Antonieta: Bien. Volviendo a mi trabajo como voluntaria del Centro de Derechos Laborales y estando en la Junta Directiva . . . Me dí cuenta que estaba organizándose una cooperativa de limpieza de casas y me decían: "Hágase miembro de la Cooperativa porque habrá trabajo y podrán luchar por que les paguen bien." Y no me atraía la idea ya que yo estaba trabajando seis días a la semana, pero como en El Salvador había yo apoyado en la organización y, luego, en la administración de una cooperativa de pequeños caficultores, y había vivido el proceso de lograr en grupo lo que individualmente no se puede, quise saber si aquí lograríamos los mismo. Así llegué a la Cooperativa y participé en los primeros talleres y me integré al Comité de Educación y aquí comenzó para mí lo más importante de ser miembro de la Cooperativa pues comencé a apoyar activamente impartiendo los talleres a nuevas miembros. Y déjeme aclarar que no se imparten sólo como un requisito, sino para darnos una mayor eficiencia en nuestras actividades pues, sabiendo el por qué de la existencia de la cooperativa, nos da seguridad para hacer un buen trabajo y desarrollarnos como personas. Como verá usted, la educación en la Cooperativa es un requisito previo y una condición permanente. Hasta aquí se ve bonito lo que es nuestra cooperativa y si usted, Drucilla, le conmovió mi caso, más me conmovió a mí conocer casos de mujeres que han pasado por verdaderos abusos y humillaciones en sus trabajos y que no son abusadas sólo por los patrones, sino por agencias y por personas de nuestra misma raza, que se dedican a negocios de limpieza y explotan sin compasión. Esta es la diferencia de la cooperativa con dichos negocios. A ellos les interesa los dólares que ganarán al conseguirle empleo a una mujer, y no les interesa saber como será tratada en el lugar de trabajo. Nuestra Cooperativa sí se interesa en que sea tratada dignamente. Por eso se le enseña a conocer sus derechos inmediatamente, que se quede o no como miembra de la Cooperativa. Lo importante es que reciban el mensaje y esto nos hace sentir que lo que estamos haciendo vale la pena y nos motiva a seguir adelante; ya que poco a poquito está cambiando el mundo. Ciertamente, me motivó para continuar en la cooperativa.

No quiero omitir que tengo una cliente de la Cooperativa (una casa), y para mí, ellos son como parte de mi familia. Tienen mi cariño y yo el de ellos. No es sólo un servicio a cambio de dinero. Yo estoy contenta trabajando para ellos, participo en sus alegrías y en sus tristezas. Y ser miembra del Centro es una gran experiencia, es una escuela. Aprendemos que organizados mejoraremos nuestra calidad de vida, y es un reto diario ya que allí encuentran apoyo todos los trabajadores inmigrantes.

D.C.: ¿Es usted feminista?

Antonieta: No.

D.C.: ¿Por qué? ¿ Es ser feminista algo malo?

Antonieta: No. Yo soy una mujer que tiene que trabajar por necesidad y no para competir con el hombre. Sí creo que tenemos los mismos derechos. Somos igual o más inteligentes que ellos.

C.M.M.: ¿Qué entiende usted por *feminista*? ¿Cuál es su definición de feminismo y de una persona feminista?

Antonieta: Bueno, feminismo es que tenemos los mismos derechos que el hombre. Pero no tenemos que tomarlo como un duelo, ni pensar que el hombre es el enemigo. Yo tengo necesidad de trabajar, sí, y lo haré para demostrar que puedo valerme por mí misma. Una mujer, de preferencia, debe dedicarse a su hogar, es el más difícil y mejor de los papeles que puede desempeñar . . . servir a la comunidad por medio de la Iglesia, Cuz Roja, etc. . . . Sin duda que estoy confundida. Por ejemplo, he escuchado a conocidas que dicen ser feministas: "Si mi esposo, o novio, se va a una discoteca con sus amigos, yo también tengo derecho a lo mismo." Y allí, yo no estoy de acuerdo, o será mi manera de pensar. Hombres y mujeres tenemos los mismos derechos, pero no los mismos privilegios.

C.M.M: Entonces, ¿no se identifica como feminista?

Antonieta: No (*riendo*). Yo no creo que me identifique como tal ¿Verdad? Simplemente como una mujer luchadora que tiene dos hijos que apoyar. Soy cabeza de familia, como muchas mujeres en este país. Lo cierto es que no sé verdaderamente lo que es el feminismo y sería interesante conocer sobre ello.

D.C.: ¿Cómo ve el futuro de *Unity*? ¿Cuál es su visión de la Cooperativa Unidad en el futuro?

Antonieta: Mi sueño es que cada vez se integren más miembras, que seamos cien o más. Pero por su conocimiento, no sólo para decir: "¡Ah! La Cooperativa Unidad del Centro de Derechos Laborales tiene cien mujeres en su membresía." No, sino que estas mujeres estarán haciendo la diferencia. Porque no aprendemos sólo cómo defendernos sino, también, a cumplir con nuestras obligaciones. Porque, si exigimos derechos, debemos cumplir con nuestras responsabilidades. Sólo unidas lograremos mejorar la calidad de vida como mujer inmigrante. El crecimiento de la Cooperativa será fantástico. Imagínese, cada una llevándole el mensaje a otra que, por alguna razón, no sea miembra. Por lo menos se les da a conocer que no tienen por qué dejarse abusar, que las leyes las protegen, que hay que buscar soluciones, que no dejen que su dignidad sea pisoteada.

D.C.: ¿La Cooperativa Unidad es parte del Centro?

Antonieta: Sí, es parte del Centro ya que aquí nació el proyecto. Nos organizaron, nos siguen apoyando con asesoría y la mayoría somos miembras del Centro, por lo que nosotras, de alguna manera, somos parte de esta gran organización sin fines lucrativos.

D.C.: Una última pregunta… ¿Existe alguna manera en que mujeres que no pertenezcan a la Cooperativa puedan ofrecer su solidaridad?

Antonieta: Ya con el hecho de estar aquí nos están apoyando. Porque les interesa el proyecto, les interesa ver de qué manera ustedes ayudan a la mujer con menos recursos, ya que esta condición—la necesidad—es la que nos hace permitir los abusos. Y si ustedes dan a conocer que somos seres humanos y merecemos respeto, esto nos ayudará a concientizar a los patrones.

D.C.: Sí.

Antonieta: Porque, igual, se están dando cuenta que, así como ellos trabajan, o usted trabaja con horario, nosotras también tenemos derecho a un horario y salario dignos que nos permita vivir dignamente.

D.C.: Antonieta, ¡muchas gracias!

Antonieta: De corazón, mi persona y la Cooperativa les agradecemos su apoyo en nuestras actividades.

Appendix C

M. E. A.
JUNIO 8, 2000
(ENTREVISTA EN ESPAÑOL)

D.C.: ¿Cuándo llegó a los Estados Unidos y de dónde es usted?

M.E.A.: Yo llegué a los Estados Unidos en 1988. Son ya doce años que estoy aquí. Soy ecuatoriana. He trabajado, la primera vez, interna, de doméstica. Pero he trabajado también en una factoría . . . diez años. Pero hace como dos años atrás, la compañía estaba que se iba, que se iba, que se iba, hasta que últimamente la vendieron y me descubrí que era mejor la cooperativa. Porque, desde que conozco la Cooperativa, yo me integré y comencé a trabajar con la cooperativa. Y . . . siempre buscando un mejor tiempo, porque soy madre de familia. Tengo tres niños. Y necesitaba tiempo para mi familia y no estar muchas horas trabajando en una factoría. En una factoría son muchas horas. Ahorita hago cuatro horas y voy temprano a recoger a mis hijos a la escuela. Tengo tiempo para estar con ellos.

D.C.: ¿Tuvo usted que hacerse cargo de niños?

M.E.A.: No, porque no había niños, pero la primera vez, cuando fui a trabajar, había mucho que hacerle, pues, a la casa. ¡Hacía años que no habían limpiado esa casa! Entonces, cuando veía que ya estaba limpia, me trajo a su mamá. Tenía que cuidar de su mamá. Aparte de lo que limpiaba la casa, me llevaba a hacer la limpieza de la oficina. Y cuando la limpieza de la oficina ya la tenía al día, bien limpia, le ayudaba a archivar. Yo hacía lo que me decía: lavar, planchar, limpieza de la casa . . .

Constanza Morales-Mair: ¿Cuánto tiempo estuvo con esta señora?

M.E.A.: Casi once meses. También me llevaba a limpiar el apartamento de su hija . . . y nunca me dieron extra en dinero. Siempre lo mismo.

D.C.: Trabajó todo el tiempo por el mismo sueldo. ¿Le importaría si le pregunto cuánto le pagaban?

M.E.A.: Me pagaban $100 a la semana.

D.C.: ¿Alguna vez le aumentaron el sueldo?

M.E.A.: No. Cuando yo le pedí que necesitaba trabajar más . . . hacer más dinero porque . . . que si podía ella recomendarme para trabajar con otra persona. Me dijo: "Tú ya puedes trabajar en otro lado porque aquí ya no te necesito." A la mamá la mandó para que regresara a Guatemala y . . . me dijo que podía buscar otro trabajo.

C.M.M.: ¿Entonces la señora era de Guatemala y la despachó?

M.E.A.: Sí. Y también me dijo que nunca dijera que yo he trabajado con ellos porque el hijo es abogado; para no perjudicar a su hijo.

C.M.M.: ¿Como amenaza?

M.E.A.: Yo no lo tomé como amenaza sino que la perjudicaría a ella porque el hijo era abogado.

D.C.: ¡Lo cual es una amenaza! Porque si no ¿cómo obtendría usted una referencia?

M.E.A.: Para mí eso era ignorancia porque . . . si me pedían referencia, no la podía dar.

D.C.: ¡Claro!

M.E.A.: Entonces una monjita de la iglesia de Nuestra Señora de Loretto me recomendó para trabajar con otra señora. Era americana. El trato era cien por ciento mejor que con la otra señora, porque me tenía mucha consideración y hasta me pagaba mucho mejor. Eran cincuenta dólares más, que era muy bueno para mí.

D.C.: Ciento cincuenta dólares.

M.E.A.: ¡Y el trabajo que hacía no era ni la cuarta parte de lo que tenía que hacer en el otro!

D.C.: ¿Cuidó niños ahí?

M.E.A.: Ahí sí había niños, pero en la tarde tenía que cuidar los niños. Por la mañana iban a la escuela. Sólo desayuno y cena tenía que preparar.

D.C.: ¿Cuánto tiempo trabajó ahí?

M.E.A.: Cinco o seis meses. La señora estaba embarazada y nació la niña. Entonces le dije si me recomendaba con otra persona pero no me pudo recomendar. Por que ella quería mandarme con una amiga pero me dijo: "Si alguien pide una recomendación y tú consigues trabajo" . . . que le dé el número de teléfono, su nombre y . . . Así lo hice cuando me salió trabajo. Pero ella quería que trabajara en el día y, en la noche, que regresara a vivir con ella. Que no pague renta.

C.M.M.: ¿Para ayudarle con los niños?

M.E.A.: ¡No! Para que yo no pagara renta. Sólo a dormir, me dijo. Pero conseguí [trabajo] en otro lado, pero interna.

D.C.: ¿No tenía que trabajar?

M.E.A.: No. Ella me dijo: "No. No te preocupes de la cena que yo aquí estoy, y hago todo el trabajo de la casa . . ." Que sólo viniera a dormir.

C.M.M.: Este nuevo era ya entonces su tercer trabajo . . .

M.E.A.: Sí. Este fue otro trabajo que he hecho interna. Ahí ya estaban los niños grandes.

C.M.M.: Los niños de la señora donde trabajó antes . . . ¿cuántos eran?

M.E.A.: Tres.

C.M.M.: ¿Y cuatro con la bebé?

M.E.A.: La bebé era la cuarta.

C.M.M.: Usted, entonces, cuidó a los tres mayores.

M.E.A.: Sí. Y cuando llegué al otro lugar que ella misma me recomendó con su referencia, donde me pagaban muy bien. Lo malo fue que decidí hacerme aquí un examen y me encontraron que yo tenía algo y me mandaron una medicina. Entonces eso, a la señora, le cayó mal . . . que la iba a contagiar a ella . . . y un montón de cosas . . . y me dijo que cuando estuviera bien regresara por el trabajo. Yo tenía una tos que . . . A veces me tocaba hacer el trabajo afuera, en el sol . . . ir a dejar a los caballos. Y adentro había demasiado aire acondicionado. Me dio una tos . . .

C.M.M.: ¿Trabajando en el campo?

M.E.A.: No, en una mansión en Hicksville, New York, aparte de aquí, retirada. Tenían caballos. Tenía que ir, dejar a los caballos donde ellos corren y comen. Ahí . . . en el potrero. Tenía que llevarlos en la mañana y traerlos a las tres de la tarde. Pero adentro había demasiado aire acondicionado y ¡me agarró una tos! Y ella me dijo que podía contagiar a ella y que me fuera a donde el doctor y que me hiciera el chequeo y, cuando yo esté bien, que traiga una carta del doctor.

D.C.: ¿Le pagó licencia por enfermedad?

M.E.A.: No. Era solamente que estuve ahí trabajando como tres semanas o quince días. No recuerdo. Ella me pagó todo pero yo conseguí trabajo en una factoría y en esa factoría estaba trabajando ya diez años y . . . yo . . . ya no quise saber más.

D.C. [*a C.M.M.*]: ¿M.E.A. tenía bronquitis y esta mujer estaba paranóica?

C.M.M [*a D.C.*]: Estaba preocupada por lo que M.E.A. pudiera tener.

D.C.: ¿Saben lo que pienso? Pienso que ella pensó que usted tenía tuberculosis. Algunas veces el racismo toma la forma de "Usted tiene una enfermedad contagiosa."

M.E.A.: Sí. Cuando me dio el médico la medicina, me dijo: "¿Tú vas a volver a trabajar ahí?" Le dije que no y me dijo: "¡Okay! ¿Tienes ya trabajo?" Le dije que sí porque cuando fui de visita sin la carta, me dijo la patrona que fuera por la carta. Cuando fui otra vez donde el médico a que me vuelva a chequear, me dijo: "Pero si sólo era una tos! ¡Cómo exagera!"

C.M.M.: ¡Un resfriado!

M.E.A.: Solamente [que era] demasiado fuerte el aire acondicionado y eso me hizo mal. Bueno. Me fui a trabajar a una factoría, mejor.

D.C.: ¿Qué tipo de fábrica?

M.E.A.: Es una factoría de . . . una empresa que es un "warehouse."

C.M.M.: ¿Y cuánto ganaba ahí?

M.E.A.: Entré ganando seis dólares la hora. Ahí había Unión y todo pero, como nunca daban aumento, yo con unas clases que aquí [en el Centro de Derechos Laborales] tuve ya, y que nos enseñaron nuestros derechos—a que tenemos que reclamar—me fui a reclamar. Pero recogimos firmas para que nos dieran una "quarter." ¡Veinticinco centavos!

C.M.M.: ¿Ese fue el aumento?

M.E.A.: Ese fue el aumento. ¡Después de ocho años!

C.M.M.: ¿Cuantas horas al día?

M.E.A.: Ocho horas al día. Cuarenta horas hacía la semana. Y a veces trabajaba "overtime" que me pagaba tiempo y medio.

C.M.M.: ¿Al cabo de cuánto tiempo dijo usted que reclamaron el aumento?

M.E.A.: Yo ya estaba como ocho años trabajando ahí.

D.C.: ¡Oh! ¡Sin aumento!

M.E.A.: Sin aumento.

D.C.: ¿Cuáles eran sus prestaciones?

M.E.A.: Teníamos "holidays"; teníamos vacaciones y seguro. Seguro familiar no tenía yo. Pero no me gustaba que la unión no hacía nada . . . que cualquier queja que uno ponía, no la escuchaban, pero cuando ellos nos iban a cobrar más—los de la unión—no nos daban un aviso ni nada, a nosotros. Solamente nos sacaban. Por ejemplo, que la Unión nos cobraba más a nosotros pero nunca nos hacían saber a nosotros. ¡Sólamente nos sacaban!

D.C.: ¿Tenían seguro y beneficios de salud?

M.E.A.: Teníamos seis días de enfermedad y, de los seis días, nos quitaron uno. Nos los dejaron en cinco y no nos dejaron saber, ni decir lo que nos habían quitado. O sea . . . este . . . Cuando iban a hacer un nuevo contrato con los trabajadores, nunca hubo . . . [alguien] que nos hablara y nos dijera: "vamos a hacer esto," o "esto nos está diciendo la compañía que vamos a hacer." Sino cuando ya habían hecho . . .

D.C.: Yo trabajé en un sindicato y fui despedida porque les dije a los trabajadores que no estaban recibiendo los beneficios que el sindicato les decía que tenían. Una unión corrupta. ¿Cuál era el nombre de su sindicato?

M.E.A.: No me recuerdo ahorita de la unión. Cuando yo le reclamé a la Unión [y pregunté] que por qué no hablaban con nosotros cuando hacían . . . cuando nos iban a quitar los "sick days," el presidente de la Unión dijo que para eso él era el representante y que él estaba autorizado para hacer todo estos cambios. "Bueno"—

le decía—"si tú eres mi representante, también deberías representarnos haciendo que nos den un aumento. Lo que tú representas es que nos pidan más producción." Porque, sí, eran altas las metas de producción que nos pedían, pero ¿aumento? ¡Nunca! ¡No nos daban más aumento!

D.C.: ¿Cuándo empezó a trabajar con las mujeres de la cooperativa?

M.E.A.: Yo estaba trabajando en la factoría y con las mujeres de la cooperativa. Yo ya trabajo dos años aquí. O sea, yo estaba trabajando los fines de semana con la Cooperativa, los sábados. Porque los "meetings" los tenemos aquí los domingos y la factoría, yo la hacía de lunes a viernes. Y los domingos, aquí, una vez al mes que nos reunimos. Para los talleres, estaba los domingos.

D.C.: ¿Qué la motivó a venir a la cooperativa?

M.E.A.: Tomé yo un curso de derechos laborales. Cuando yo tomé eso, entonces comencé mi trabajo en la factoría ayudando también a que las demás conozcan sus derechos: que tenemos que reclamar a la Unión; que no nos dé ese trato—nos gritaban que eramos "Spanish . . . mierda"—el "boss." Porque le decíamos que cómo es que nos exige que hagamos unas órdenes demasiado pesadas. Nosotras hacemos las órdenes pero que pongan a un hombre a que saque las cajas que eran muy pesadas. Entonces el boss era cuando usaba esas palabras tan . . . ¡sucias! Verá . . . Es que las mujeres recogíamos la mercancía y la organizábamos. Por ejemplo: esto era un "sheet" que estaba lleno de órdenes que debería sacar: perfumes, medicinas . . . lo que pida la cliente, y una tenía que recoger todo eso. Y todo eso era demasiado pesado porque van hasta ollas y cosas de metal pesado . . . sillas . . . ¡demasiado pesado! Pedíamos que trajeran a un hombre y el "boss" no quería. Entonces, hubo una muchacha que se enojó con el "boss." Lo que ocurrió fue que esta muchacha no puso la mercancía pesada y vino él y, entonces, le quitó los "labels" y le dijo: "¡Véte para tu casa!" Entonces, como yo ya estaba más o menos enseñada sobre derechos y sabía que él estaba haciendo una injusticia con nosotras . . . Yo le había comentado a los compañeros que teníamos derecho a enfrentarnos al "boss," así es que, la siguiente vez que pasó un incidente, nos negamos a seguir trabajando por el día.

D.C.: ¿Se considera usted feminista?

M.E.A.: Sí. Soy feminista.

C.M.M.: ¿Podría hablarnos de nuevo sobre lo que nos contó antes . . . de cómo, habiendo señoras mayores entre las trabajadoras, usted se interesó por el bienestar de todas?

M.E.A.: Bueno, pues había señoras que tenían más de cuarenta o cincuenta años y que habían trabajado de veinte a veinticuatro años ahí. Bueno, yo estaba ahí diez años, pero yo me sentía todavía joven, pero mis energías estaban acabadas. Llegaba tensa y nerviosa . . . que ya no podía trabajar fuerte. Entonces yo les digo: "¿Cómo es que nos quieren seguir pidiendo más producción si nuestras energías ya no dan más? Y esas señoras ¡ya no daban más! Eran señoras de otra raza, haitianas e hispanas. Y también estaban diecinueve muchachas en la lista que tenía el jefe y que las iba a llevar a la oficina a que firmaran un documento, así uno no lo acepte. Pero el supervisor decía que tenían que firmar, y si uno se negaba, lo firmaba el supervisor [por ellas] haciendo constar que la trabajadora lo había firmado. Y eso era injusto. Porque yo estoy aceptando que no puedo hacer más, mi trabajo. Por más que trataba, no, no lo podía hacer. Y obligaban a las señoras a hacer eso. Ahí es cuando comenzamos a mobilizarnos. Porque eso estaba siendo demasiada explotación.

D.C.: M.E.A., . . . Una pregunta más. ¿Qué espera para la niña . . . para su hija?

M.E.A.: No sólo para mi hija. Yo lo desearía para muchos niños más. Que en realidad estudien, se preparen. Porque si uno tiene un título, de algo, uno puede trabajar donde quiera. Porque, lastimosamente, yo no tuve esa oportunidad, porque mis padres eran bien pobres. Y yo busco eso para mis hijos. Algo. Un futuro mejor donde puedan ganarse la vida.

D.C.: Otra . . . ¡Otra pregunta más! ¿De donde saca usted su valentía? ¡Yo sé lo difícil que es enfrentarse al jefe!

M.E.A.: Sí. Por peleona y reclamar mis derechos he perdido quizás algo. Ahorita necesito una carta para llenar . . . para la tarjeta verde, la "green card," porque eso piden en inmigración. Yo pedí eso al trabajo. Como yo hice todo esto, me dijeron que me la iban a mandar por correo, y hasta el día de hoy no me ha llegado. Me dijeron que, como yo peleé con el jefe y con la Unión, esa fue la razón por la que no me dieron la carta. Y yo fui con un compañero, y los demás vieron. Eramos dos . . . A mí no me ha llegado la carta. Yo necesitaba esa carta. Yo creo que sentían que yo estaba haciendo algo por la justicia . . . por hacer que respetaran a los trabajadores. Si no la envían . . . digo . . . bueno . . . de algun lado la voy a obtener. Voy a ir ahorita—esta semana quizás—antes de que cierren, donde está la oficina principal de la compañía, en Westbury, New York, a pedirla a la que está a cargo de todos los trabajadores. Porque yo le pedí al de la Unión y no me la ha mandado. Le pedí al supervisor ahí . . . al encargado de la fábrica, y me dijo que la Unión va a mandármela. Por haber contestado a veces me habían dicho—mi supervisora—que yo era atrevida por decir: "¿Por qué nos pones demasiada producción? Si no soy máquina para sacarla! Yo no puedo hacer milagros." Y había puesto quejas; que era que yo no quería hacer el trabajo. Hasta ahí me llegaron, una vez, a eliminarme. Pero me ganaba a otros supervisores y me entraba a otro departamento a seguir trabajando con otros y hasta que la otra veía que estaba trabajando de regreso y ella me había eliminado . . . Todo esto por reclamar nuestros derechos. Ya no trabajo más con la factoría y me siento feliz trabajando con la *Cooperativa de Limpieza de Casas Unidad*, porque tengo más tiempo con mis hijos. Estoy aprendiendo inglés y muchas cosas más.

NOTES

1. One of the most important discussions of this issue took place in 1997. It was part of the now famous Philosophers' Brief, an amicus curiae brief filed in two Supreme Court cases that were argued during that year. (See Ronald Dworkin, Thomas Nagel, Robert Nozick, John Rawls, Thomas Scanlon, and Judith Jarvis Thomson, "Assisted Suicide: The Philosophers' Brief," *New York Review of Books* [March 27, 1997].) The ultimate conclusion of these six eminent moral and political philosophers was that, as a matter of constitutional law, "Each individual has a right to make 'the most intimate and personal choices central to personal dignity and autonomy.' That right encompasses the right to exercise some control over the time and manner of one's death." As John Rawls himself later explained, "We wanted the Court to decide the cases in terms of what we thought was a basic constitutional right. That's not a matter of religious right, one way or the other; it's a constitutional principle. It's said to be part of American liberties that you should be able to decide these fundamental questions as a free citizen. Of course, we know that not everyone agrees with assisted suicide, but people might agree that one has the right to it, even if they're not themselves going to exercise it." (See John Rawls, "*Commonweal* Interview with John Rawls," *Collected Papers*, ed. Samuel Freeman [Cambridge, MA: Harvard University Press, 1999], p. 618.) Against this position, Michael Walzer has argued that because too few citizens have the means to exercise their freedom through the law, the right to physician-assisted suicide cannot actually protect all persons, thereby rendering it exclusionary and hierarchical. (See Michael Walzer, "Feed the Face," *The New Republic* [June 9, 1997].)

   Philosophers working outside the fields of analytic jurisprudence and Anglo-American legal theory have also debated the vexing issue at the level of personal autonomy and self-determination. Richard W. Momeyer, for instance, thinks that if a person loses the ability to make decisions for herself, others have to act on her behalf according to these criteria: *autonomy* (What would this person choose for himself or herself if he or she could choose?), *best interest* (What is the best interest of this person?), and *moral duty* (What ought this person to do if in a position to make moral choices?). (See Richard W. Momeyer, *Confronting Death* [Bloomington and Indianapolis, IN: Indiana University Press, 1988], p. 150.) Norman L. Cantor, however, offers a rather different vision of the person who enters "post-competency": "Self-determination in shaping medical intervention during any naturally occurring dying process upholds personal priorities and thus honors human capacity for choice. When that self-determination is upheld at a post-competence stage of existence, society honors the *fulfillment* of human capacity for choice as represented in each person's shaping of his or her lifetime priorities and character. In prescribing post-competence care, every declarant seeks to preserve a personal vision of dignity, to imprint his or her own character on this critical juncture and to continue personal ideals." (See Norman L. Cantor, *Advance Directives and the Pursuit of Death with Dignity* [Bloomington and Indianapolis, IN: Indiana University Press, 1993], p. 107.) Other key debates have been waged around the very concept of person and the purported sanctity of human life as against other forms of life. (See Michael Tooley, "Decisions to Terminate Life and the Concept of Person" and Peter Singer, "Unsanctifying Life," in *Ethical Issues Relating to Life and Death*, ed. John Ladd [Oxford: Oxford University Press, 1979], p. 62–93; 41–61). For a good general history of the right to die, see Derek Humphrey and Ann Wickett, *The Right to Die: Understanding Euthanasia* (New York: Harper & Row, 1986).

2. The Western philosophical understanding of dignity takes its perhaps fullest shape in the thought of Immanuel Kant. As a Kantian concept, dignity has infinite meaning because it follows from the grandeur of every person acting as a universally self-legislating, rational being in an ideal pursuit of human freedom whose realization is always to come. In Kant's rightly famous words, everyone is "free with respect to all laws of nature, obeying only those which he himself gives and in accordance with which his maxims can belong to a giving of universal law (to which at the same time he subjects himself). For, nothing can have a worth other than that which the law determines for it. But the lawgiving itself, which determines all worth, must for that very reason have a dignity, that is, an unconditional, incomparable worth . . . the dignity of humanity consists just in this capacity to give universal law, though with the condition of also being subject to this very lawgiving" (See Immanuel Kant, *The Groundwork of The Metaphysics of Morals*, in *Practical Philosophy*, ed. Mary J. Gregor, The Cambridge Edition of the Works of Immanuel Kant [Cambridge: Cambridge University Press, 1996], pp. 85, 89). In an equally important formulation, Kant declares that "a human being regarded as a *person* . . . is exalted above any price; for as a person . . . he is not to be valued merely as a means to the ends of others or even to his own ends, but as an end in itself, that is, he possesses a *dignity* (absolute inner worth) by which he exacts *respect* for himself from all other rational beings in the world. He can measure himself with every other being of this kind and value himself on a footing of equality with them." (See Immanuel Kant, *The Metaphysics of Morals*, in *Practical Philosophy*, p. 557.) Long before contemporary liberal political philosophers began to spar over the question of whether to privilege equality or freedom, Kant showed there was absolutely no reason for such a dispute. His philosophically radical conception of freedom says that we are equal precisely *as* free beings. To argue the reverse—that we are free *only* by means of equality—is to deny the central role dignity plays in allowing us to witness the immeasurable fact that we shall all remain *equally free*. Jean-Luc Nancy understands the sheer immensity of thinking our freedom which, for everyone, is simply *that* free: "We are . . . not free to think freedom or not to think it, but thinking (that is, the human being) is free *for* freedom: it is given over to and delivered for what from the beginning exceeded it, outran it, and overflowed it. But it is in this way that thinking definitively keeps its place in the world of our most concrete and living relations, of our most urgent and serious decisions." (See Jean-Luc Nancy, *The Experience of Freedom* [Stanford: Stanford University Press, 1993], p. 8).

3. Zora Neale Hurston, *Their Eyes Were Watching God* (New York: HarperCollins Publishers, 1990), p. 16.

4. As I have suggested elsewhere, "The idea of humanity *as an idea*, Kant reminds us, is contentless. Humanity, indeed, is just one example of the postulation of free creatures with the capacity of shaping their own destiny." (See Drucilla Cornell, "Antiracism, Multiculturalism, and the Ethics of Identification," *Just Cause: Freedom, Identity, and Rights* [Lanham, MD: Rowman & Littlefield, 2000], p. 36.) Nevertheless, for Kant, it is quite paradoxical that "the mere dignity of humanity . . . without any other end or advantage to be attained by it— *hence respect for a mere idea*—is yet to serve as an inflexible precept of the will, and that it is just in this independence of maxims from all such incentives that their sublimity consists, and the worthiness of every rational subject to be a lawgiving member in the kingdom of ends; for otherwise he would have to be represented only as subject to the natural law of his needs. . . . Our own will insofar as it would act only under the condition of a possible giving of universal law through its maxims—*this will possible for us in idea*—is the proper object of respect." (See Immanuel Kant, *Groundwork of The Metaphysics of Morals*, in *Practical Philosophy*, p. 88, emphasis mine.) The pure ideality of humanity in its dignity gives human beings their freedom. It is the full autonomy of will, the freedom simply to be—without being determined by the causality of nature— that defines the sensibility of the world. If the idea of humanity possessed ideational content, it would *always* be directed toward thoughts of objects, ends, and advantages not yet present to the senses. Even purely at the level of ideas, then, it would *already* be unfree. It can remain free only if—and this, precisely, goes to the heart of Kant's *paradox*—its ideality is completely without ideation. When we think of our freedom as irreducible to the limits of the empirical world, the idea of humanity endlessly frees itself. (This is crucial for understanding why it does not make philosophical sense to try—as many still do—to hold freedom to a strictly ontological horizon. Freedom will continue to thwart each and every containment policy of *presence*.) Human beings are thus eminently free to think their very limitlessness, their freedom, which they will realize as members of an *ideal* kingdom of ends. It is a matter of the setting, determination, mission, goal, destination—in a word, vocation—*for* freedom; all of which may be expressed with the German word *Bestimmung*. Yet, in the end, it is also a matter of recognizing the way in which this ideal of freedom consists in the permanent radicalization of the idea of humanity, thereby collapsing the theoretical distance between them. "Any high praise for *the ideal of humanity* in its moral perfection," writes Kant, "can lose nothing in practical reality from examples to the contrary, drawn from what human beings now are, have become, or will presumably become in the future . . ." (See Immanuel Kant, *The Metaphysics of Morals*, in *Practical Philosophy*, p. 534, emphasis mine.)

5. See my book, *At the Heart of Freedom: Feminism, Sex, and Equality* (Princeton: Princeton University Press, 1998), pp. 6–8.

6. Alicia Ostriker, "The War of Men and Women," *The Imaginary Lover* (Pittsburgh: University of Pittsburgh Press, 1986), p. 79.

7. With the recognition that it is always a being with history, finitude, and limits— not some phantasmic object—we have lost, mourning becomes not simply an act of remembering the other but a long, difficult process of being called to do justice *through* remembrance. My understanding of mourning has much more

in common with Jacques Derrida's work on the subject than with contemporary efforts to rethink the Freudian relation between mourning and melancholia. Shortly after the death of Paul de Man, his beloved friend and intellectual compatriot, Derrida famously asked: "What is an impossible mourning? What does it tell us, this impossible mourning, about an essence of memory? And as concerns the other in us, even in this 'distant premonition of the other,' where is the most unjust betrayal? Is the most distressing, or even the most deadly infidelity that of a possible mourning which would interiorize within us the image, idol, or ideal of the other who is dead and lives only in us? Or is it that of the impossible mourning, which, leaving the other his alterity, respecting thus his infinite remove, either refuses to take or is incapable of taking the other within oneself, as in the tomb or the vault of some narcissism?" (See Jacques Derrida, *Memoires: For Paul de Man* [New York: Columbia University Press, 1986], p. 7.)

The narcissistic impulse toward interiorization Derrida amplifies here is precisely what informs Freud's analysis of mourning and melancholia. As Giorgio Agamben explains, " . . . the dynamic mechanism of melancholy borrows its essential characteristics from mourning and in part from narcissistic regression. As when, in mourning, the libido reacts to proof of the fact that the loved one has ceased to exist, fixating itself on every memory and object formerly linked to the loved object, so melancholy is also a reaction to the loss of a loved object; however, contrary to what might be expected, such loss is not followed by a transfer of libido to another object, but rather by its withdrawal into the ego, narcissistically identified with the lost object." (See Giorgio Agamben, *Stanzas: Word and Phantasm in Western Culture* [Minneapolis: University of Minnesota Press, 1993], p. 19.) Melancholy is pathological in the Freudian sense because whereas "mourning follows a loss that has really occurred, in melancholia not only is it unclear what object has been lost, it is uncertain that one can speak of a loss at all" (20). Yet, as Agamben goes on to suggest, melancholy is unstable enough as a pathology to allow for at least two different libidinal experiences. The first involves fetishism: "the object is neither appropriated nor lost, but both possessed and lost at the same time. And as the fetish is at once the sign of something and its absence, and owes to this contradiction its own phantomatic status, so the object of the melancholic project is at once real and unreal, incorporated and lost, affirmed and denied" (21). The second involves privative nostalgia, a strangely romantic dialectic of death and redemption: "melancholy would be not so much the regressive reaction to the loss of the love object as the imaginative capacity to make an unobtainable object appear as if lost. If the libido behaves *as if* a loss had occurred although *nothing* in fact has been lost, this is because the libido stages a simulation where what cannot be lost because it has never been possessed appears as lost, and what could never be possessed because it had never perhaps existed may be appropriated insofar as it is lost . . . the strategy of melancholy opens a space for the existence of the unreal and marks out a scene in which the ego may enter into relation with it and attempt an appropriation such as no other possession could rival and no loss possibly threaten" (20).

When it is a matter of respecting the life as well as the death of a loved one, it hardly seems just to embrace a psychoanalytic triumphalism that crudely uplifts subjectivity rooted in melancholy only at the expense of diminishing the very possibility of mourning. This is the mistake of much contemporary theory—a mistake Slavoj Žižek, perhaps not surprisingly, has taken great pleasure in caricaturing. In characteristically overwrought, overconfident fashion, Žižek exclaims that, against Freud, what we are supposed to do these days is "assert the conceptual *and* ethical primacy of melancholy. In the process of the loss, there is always a remainder that cannot be integrated through the work of mourning, and the ultimate fidelity is the fidelity to this remainder. Mourning is a kind of betrayal, the second killing of the (lost) object, while the melancholic subject remains faithful to the lost object, refusing to renounce his or her attachment to it." (See Slavoj Žižek, "Melancholy and the Act," *Critical Inquiry* 26:4 [Summer 2000], p. 658.)

As should be abundantly clear from Agamben's account of Freud, the melancholic subject has a very troubled, obsessive relationship with its object of loss, a lost possession whose "actual existence" was only as part of a phantasmally real universe. Its purported faithfulness to its object should thus be defined more precisely as a kind of "bad faith" or disingenuous fidelity to an object that may or may not have ever existed. In this way, it makes little sense—and indeed may be deeply unethical—to defend an ethics of melancholy whenever real people—beings, not just things or objects (both of which Heidegger rightfully consigned to the ontic realm of "entities")—are at stake. We must instead assert the primacy and perhaps even the purely originary status of mourning as a structure of experience that remains bigger than us all, that precedes and surpasses us as we remember it, recall it, live it, during the short time anyone of us is alive:

> We weep *precisely* over what happens to us when everything is entrusted to the sole memory that is "in me" or "in us." But we must also recall, in another turn of memory, that the "within me" and the "within us" *do not* arise or appear *before* this terrible experience. Or at least not before its possibility, actually felt and inscribed in us, signed. The "within me" and the "within us" acquire their sense and their bearing only by carrying within themselves the death and the memory of the other; of an other who is greater than them, greater than what they or we can bear, carry, or comprehend, since we then lament being no more than "memory," "in memory." Which is another way of remaining inconsolable before the finitude of memory. We know, we knew, *we remember*—before the death of the loved one—that being-in-me or being-in-us is constituted out of the possibility of mourning. We are only ourselves from the perspective of this knowledge which is older than ourselves; and this is why I say that we begin by *recalling* this to ourselves: we come to ourselves through this memory of *possible* mourning. . . . Upon the death of the other we are given to memory, and thus to interiorization, since the other, outside us, is now nothing. And with the dark light of this nothing, we learn that the other resists the closure of our interiorizing memory. With the nothing of this irrevocable absence, the other appears *as other*, and *as other for us*, upon his death or at least in the anticipated possibility of a death, since death constitutes and makes manifest the limits of a *me* or an *us* who are obliged to harbor something that is greater and other than them; *something outside of them within them*. (*Memoires: For Paul de Man*, pp. 33–34)

8. See Drucilla Cornell, *The Imaginary Domain: Abortion, Pornography, and Sexual Harassment* (London: Routledge, 1995).

CHAPTER 1

1. Later on in my life, I was able to make the distinction between harming someone and wronging someone. See Immanuel Kant, "On A Supposed Right to Lie Because of Philanthropic Concerns," *Ethical Philosophy* (Indianapolis: Hackett Publishing Company, 1994), pp. 162–166.

2. In *Their Eyes Were Watching God*, Zora Neale Hurston often uses the imagery of the sun's rays to illuminate each of the characters' dignity, specifically Nanny's.

3. Within the Western liberal traditions, dignity is often associated with individualism. We have dignity because we are autonomous and understood as separate from other people, and self-determining with respect to our destinies. This is why we are ends-in-ourselves and cannot be reduced to a mere means to someone else's ends, even if those ends are encompassed in a vision of the good life. Or so the charge of an inaccurate conception of actual human beings has been made against liberals, including some who defend something close to the ideal of dignity as I am defining it. This understanding of autonomy has rightly taken critical heat from a number of different philosophical positions and political perspectives. I accept that critique—as do many liberals—and yet I will insist that dignity need not be integrally connected to individualism so understood. See generally, Michael J. Sandel, *Liberalism and the Limits of Justice* (Cambridge: Cambridge University Press, 1998).

4. See Uday Singh Mehta, *Liberalism and Empire: A Study in Nineteenth Century British Liberal Thought* (Chicago: University of Chicago Press, 1999).

5. See *Fired for Crying to the Gringos: The Women in El Salvador Who Sew Liz Claiborne Garments Speak Out Asking for Justice* (New York: National Labor Committee, July 1999); Nicole L. Grimm, "The North American Free Trade Agreement on Labor Cooperation and Its Effects on Women Working in Mexican Maquiladoras," *American University Law Review* 48 (October 1998): pp. 179–227.

6. See Saskia Sassen, "Notes on the Incorporation of Third World Women Into Wage Labor Through Immigration and Offshore Production," *Globalization and Its Discontents* (New York: The New Press, 1998), pp. 111–131.

7. For detailed discussions of how poorly women fare world wide in conditions of inequality, see *Women, Culture, and Development: A Study of Human Capabilities*, eds. Martha C. Nussbaum and Jonathan Glover (New York: Oxford University Press, 1995).

8. See Shulamith Shahar, *The Fourth Estate: A History of Women in the Middle Ages*, trans. Chaya Galai (London: Methuen & Co., 1983), pp. 95–97.

9. Manuscript on file with the author.

10. Phone interview with Edward Bergstrom, March 2001.

11. See, for example, Catharine A. Mackinnon, "The Roar on the Other Side of Silence," *In Harm's Way: The Pornography Civil Rights Hearings* (Cambridge, MA: Harvard University Press, 1997), pp. 3–24.

12. See Joan Wallach Scott, *Only Paradoxes to Offer: French Feminists and the Rights of Man* (Cambridge, MA: Harvard University Press, 1996). Scott analyzes the lives of four feminists who fought for woman's suffrage and equality during different revolutionary moments in French history from 1789 to 1944. She argues that "it was in moments of revolution or constitutional transformation

that the question of political rights was most open to discussion" (14). All of these women paid high prices for insisting that the principles of equality should be applied to women.

13. I borrow this phrase from the title of Marie Cardinal's book, *The Words to Say It*, which will be discussed at length in chapter 2. See Marie Cardinal, *The Words to Say It* (Cambridge, MA: Van Vactor & Goodheart, 1975).

14. For what I mean by "beyond," see Drucilla Cornell, "The Maternal and the Feminine: Social Reality, Fantasy and Ethical Relation," *Beyond Accommodation: Ethical Feminism, Deconstruction, and the Law* (Lanham, MD: Rowman & Littlefield, 1999), pp. 21–78.

15. This emphasis on the social construction of femininity is shared by almost all feminists of the "second wave." Despite her deep ambiguity about femininity, it was de Beauvoir herself, rightfully credited with being one of the mothers of the movement, who painstakingly showed us how femininity was symbolically and socially constructed as a set of limits on who a woman could be. See generally, Simone de Beauvoir, *The Second Sex*, trans. H. M. Parshley (New York: Vintage Books, 1974).

16. It was texts like Simone de Beauvoir's pathbreaking work *The Second Sex* and in this country, Betty Friedan's *The Feminine Mystique* that first exposed femininity as a set of encoded rules and norms for behavior. It was in the rebellion against them that they were made visible for conscious examination. This conscious examination of imposed norms as the shaped and limited possibilities of women's lives became a collective process known as consciousness raising.

17. See Sheila Tobias, *Faces of Feminism: An Activist's Reflections on the Women's Movement* (Boulder, Colorado: Westview Press, 1997). Tobias argues that as feminist activity in the United States declined after the ratification of the Nineteenth Amendment, "equal rights feminism lost credibility as a progressive movement among the working class, social reformists, and younger women" (40).

18. See generally Susan Brownmiller, *Femininity* (New York: Linden Press/Simon & Schuster, 1984). Brownmiller examines the historical development of an aesthetic of femininity as a set of imposed limitations on women that function, among other things, "as a value system of niceness" (17).

19. Brownmiller, *Femininity*, p. 17.

20. See, for example, Stephanie Coontz, *The Way We Never Were: American Families and the Nostalgia Trap* (New York: Basic Books, 1992), particularly chapter 3, "'My Mother Was a Saint': Individualism, Gender Myths, and the Problem of Love."

21. See John Church, *Pasenda Cowboy: Growing Up in Southern California and Montana 1925 to 1947* (Novato: Conover-Patterson, 1996), p. 137.

22. Warren Bradford Cornell, my brother, is a professor of finance and the director of the Bank of America Research Center at UCLA's Anderson School of Management. He was the founder and former president of Finecon, which was sold in 1998 to the Charles River Associates. He is now the senior consultant and testifying witness at Charles River Associates.

23. To quote the rightfully famous work of Nicolas Abraham and Maria Torok: "The father's family romance was a repressed fantasy: the initially restrained and finally delirious preoccupation of the patient seems to be the effect of being haunted by a phantom, itself due to the tomb enclosed within his father's

psyche." See Nicolas Abraham and Maria Torok, "Notes on the Phantom: A Complement to Freud's Metapsychology," *The Shell and Kernel: Renewals of Psychoanalysis* (Chicago: University of Chicago Press, 1994), p. 173.

24. See Drucilla Cornell, *The Imaginary Domain: Abortion, Pornography and Sexual Harassment* (New York: Routledge, 1995); *At the Heart of Freedom: Feminism, Sex, and Equality* (Princeton: Princeton University Press, 1998).

25. Immanuel Kant described the feeling of respect as generated by the experience of the sublime: "The feeling of our incapacity to attain to an idea that is law for us, is respect." (See Immanuel Kant, *Critique of Judgment*, trans. James Creed Meredith [Oxford: Oxford University Press, 1952], p. 105.) And yet, when we exercise our moral freedom, even if we can never grasp it through our theoretical knowledge, we confront the greatness of our humanity in the face of our inevitable succumbing to our natural side. Death is the fate of every human being. Since he thought it controverted the dignity of our humanity, Kant himself argued against suicide. I disagree and cannot but believe that Kant himself would have come around had he lived to confront recent debates about the right to die. Suicide need not be an exercise of our moral freedom, but I witnessed that it can be.

NOTES TO CHAPTER 2

1. At first it may seem odd to refer to respect as a feeling. We are more used to hearing about respect in political philosophy as an attitude often described as a concern that the state should show its citizens and that citizens should show each other in their public life. (See Ronald Dworkin, "Liberalism," *A Matter of Principle* [Cambridge, MA: Harvard University Press, 1985], pp. 181–204.) Dworkin writes that one of the principles of equality, as a political ideal, requires that "the government [must] treat all those in its charge *as equals*, that is, entitled to its equal concern and respect" (190). Respect as a feeling is evoked by a confrontation with a sublime object. I am following Kant in his discussion of the feeling of respect when confronted by a sublime object including the evocation of our inscrutable freedom as moral beings. As is well known, Kant himself was opposed to suicide. (See "Man's Duty to Himself as an Animal Being," *The Metaphysics of Morals*, trans. Mary Gregor (Cambridge: Cambridge University Press, 1991), p. 281.) In chapter 3, in my discussion of Gayatri Spivak's recent work, *A Critique of Postcolonial Reason: Toward a History of the Vanishing Present* (Cambridge: Harvard University Press, 1999), I will return to the discussion of women's suicide presented to us as a sublime object. In this chapter, I focus less on my central disagreement with Kant—that there are circumstances when suicide is an exercise of moral freedom. Obviously, this is no longer an abstract question for me since my mother wanted me to agree with her decision and to understand her act as one of moral freedom. So I will instead examine how Spivak evokes several different kinds of suicide and the circumstances in which they were committed, or in some cases not committed, as aesthetic examples of the *pathetically sublime*, a phrase I borrow from Friedrich Schiller. (See "On the Pathetic," in *Friedrich Schiller: Essays*, eds. Walter Hinderer and Daniel O. Dahlstrom, trans. Daniel O. Dahlstrom [New York: Continuum Publishing Company, 1998], pp. 45–69.) As I will argue, this evocation

can be suicide only if the dignity of the woman involved is postulated and projected onto the representation of her and then onto the representation un-
folded in Spivak's narrative. See my chapter 3.

2. See Trinh T. Minh-ha, "Grandma's Story," *Woman, Native, Other: Writing
    Postcoloniality and Feminism* (Bloomington: Indiana University Press, 1989), pp.
    121–122.

3. There are, of course, other reasons for turning to psychoanalysis in feminist
    theory, to say nothing of political and social theory. For example, psychoanaly-
    sis brings depth to the critique of the individualist conception of the subject or
    self. We are shaped as much by what has gone before us as we are by the pres-
    ent social world we inhabit. In this way, psychoanalysis offers us a critique of
    the superficial understandings of the social construction of the subject that do
    not see how history lives on within us. It thus actually allows us to distance
    ourselves from any purely social self.

4. I need to explain why I write about psychoanalysis so generally. Disagree-
    ments between different schools in psychoanalysis are often turned into fero-
    cious battles in which you have to be either on one side or the other. But to
    take sides—that is exactly what I am not going to do in this chapter. Not sim-
    ply because I find battles of this sort unproductive; nor because I am an out-
    sider by not being a clinician. Instead, it has to do with the role I give to
    psychoanalysis in political and social theory. The lessons we draw from psy-
    choanalysis must be broad enough to be accepted by a number of different
    schools of psychoanalysis. If political philosophy and social theory are to play
    some role in advocating ethical and political ideals as well as actual programs
    of reform—and I clearly think they can—we cannot hope to premise solidar-
    ity on these reforms by insisting that full agreement be reached with the
    tenets of any one school of psychoanalysis concerning questions of the sub-
    ject, the self, and so on. See, for example, my argument in favor of the ideal of
    the imaginary domain, the moral and psychic space we need to have protected
    in order to work through our sexual identifications (*The Imaginary Domain:
    Abortion, Pornography, and Sexual Harassment* [New York: Routledge, 1995];
    At The Heart of Freedom [Princeton: Princeton University Press, 1998]). In a
    sense, then, my project involves exploring what different schools share, par-
    ticularly when we think about the specific relationship between feminist-in-
    spired contributions to psychoanalysis and their relationship to feminist
    theory more generally.

    True to that effort, I will focus on the writings of several thinkers, some of
    whom identify themselves with the French analyst and theorist Jacques Lacan,
    others who associate with object relations or intersubjective theory as it has
    primarily been developed in the United States. This may seem an unlikely, if
    not impossible, combination. Lacanian theory emphasizes the necessity for
    symbolic castration—the psychic separation of the child from its primordial
    Others as the ultimate psychic law to which we must be subjected if we are to
    undergo the birth of our desire and the formation of subjectivity. Sexual differ-
    ence is analyzed in its structural relationship to the law of symbolic castration,
    which, in many well-received readings of Lacan, has the effect of rendering the
    feminine as that which cannot be represented. I will address why this is the
    case and why I reject the supposedly integral relationship between symbolic

castration and the impossibility of representing the feminine in language. (For an earlier discussion of mine concerning why I disagree with some Lacanians about the unrepresentability of the feminine within sexual difference, see Drucilla Cornell, "The Maternal and the Feminine," *Beyond Accommodation: Ethical Feminism, Deconstruction, and the Law* [Lanham, MD: Rowman & Littlefield, 1999], pp. 21–78.)

Object relations theory and intersubjective theory, on the other hand, begin with the actual relationships between parents and children. Feminist object relation theorists analyze the gender divide through the object relation to the mother. The best example is, of course, Nancy Chodorow's pathbreaking book, *The Reproduction of Mothering: Psychoanalysis and the Sociology of Gender* (Berkeley: University of California Press, 1978). The foundational assumption is that the mother as well as the child's relationship to her can be represented. How the mother is represented will matter a great deal both in the lives of actual children and in how women's gender identity can be understood and transformed. At first glance, the differences between these theoretical and analytical approaches may seem insurmountable. Without glossing over the significance of those differences, I want to argue that even though they begin from divergent analytic positions, some thinkers in both camps end up with a similar emphasis on the ethical conditions out of which desire and subjectivity can arise. My choice of theorists has to do with my interpretation of them as fellow travelers in the effort to defend dignity as the ethical basis for certain psychic laws in intergenerational relationships and for their insistence on the importance of women claiming their desire in their struggle for freedom.

5. Even if one agrees with this insight, one might still reject psychoanalysis. Indeed, the question arises: does psychoanalysis have any real place in feminist political and social theory, both of which are usually defined as having public affairs and social conditions as their object of inquiry? Politically committed theorists who are not feminists but who are sympathetic to psychoanalysis—particularly to Sigmund Freud's discovery of the unconscious as it spills over into our lives—have argued that it is helpful in illuminating our private fantasies and our particular emotional dramas. But that would be the exclusive service it can provide us. The very complexity of the human psyche, which psychoanalysis demands us to confront, plays out in the fantasies that also make us unique individuals. Its attention to the depth of any individual's psychic life makes it a poor tool for generalizations about our social conditions or for the advocacy of political reforms that would reshape our shared public life. Or so the story goes. Frequently, thinkers who make this argument about the limits of psychoanalysis when used in the service of political theory explicitly appeal to the importance of the division between public and private life and the need to maintain the liberal values dependent on that division.

Many feminist theorists have challenged this division outright or at least questioned its misuse in a social order that is patriarchal. By *patriarchy*, I mean both patrilineal lineage and the masculine privileges that have been associated with it. Most feminists of my generation have used the word broadly to designate a society of male domination which conflates the masculine with the human as a result of such domination. Throughout the 1970s, feminists worked hard to show that the so-called private realm of home and family was not truly private since it was heavily regulated by a law enforcing certain moral restrictions on what could count as a legitimate family. To the degree that families were actually left alone by the law, it was also often at the detriment of women and their children. The slogan "the personal is political" meant to challenge the reigning idea that our private and intimate lives were ever completely our own—that is to say, truly private—in a patriarchal society. There have been many stories and interpretations of what that slogan actually means and how psychic lives are profoundly influenced by the political, social, and symbolic order in which we live. For my purposes here, I use the slogan simply to point out that the feminist suspicion of psychoanalysis, which certainly continues to exist, is not motivated by the idea that the traditional liberal divide between the public and the private places limits on the use-value of psychoanalysis in social theory and political philosophy. Even so, the suspicion is still there.

Their wariness of some of the central tenets of psychoanalysis has led some feminists to reject its claim upon us altogether. Yet feminists who reject psychoanalysis do not all make that move for the same reasons. It is a risky business to try to say what these critiques share if they do, indeed, share anything at all. Still, one seemingly shared line of fire is directed toward its patriarchal assumptions about the appropriate identities for mature men and women to assume. Of course, feminist psychoanalysts throughout the last century have also been at the forefront in battling against certain schools of psychoanalysis that purportedly sought to enforce "normalized" heterosexual gender identities as crucial to the mental health of human beings. But the presence of feminists within the profession and the critiques they have offered have not lifted the suspicion among some feminists outside psychoanalysis. It is no exaggeration to say that some of the classic works of the second wave of feminism made psychoanalysis one of its main targets. Penis envy was belittled because to the degree that such envy existed, it had nothing to do with penises. It did not refer to the actual body part, but to the potency it represented. Who would not want to be a boy when boys were the ones born to freedom with the promise of power, including over women in their future as men? Hysteria, then, was the righteous refusal to settle into the prison of femininity. All the ills of civilization seemed to be placed on the shoulders of women, particularly mothers; analytic insight was targeted for reducing itself into an endless series of lessons about how mothers might do better, particularly for their sons, since there was not much to be done for their daughters anyway, fated as they were to be girls. In short, if women were "mad," they had the right to be. Psychoanalysts who wanted to cure them of their madness by teaching them how to accept the confinements of femininity were part of the problem, not the solution.

Feminists of color specifically critiqued the careless generalizations made about human nature, sexuality, gender, and the family that, far from being universal, are actually reflections of Western culture at a particular moment in its history. Again, psychoanalysis was accused of being on the wrong side ethically and politically, but this time in the battle against racism and imperialism. A classic example of being on the wrong side was the way in which some schools in psychoanalysis found African American families to have a pathological structure.

Is psychoanalysis inevitably implicated in wrongly taking insight into mainly straight, white upper-middle-class families with traditional gender identities, and offering such insight in the form of universal truths about the human condition particularly as it is expressed in sexuality, love relationships, and child rearing?

These are serious questions about whether psychoanalytic theory has a blind spot at its very conceptual core. Obviously, these questions need to be addressed. Given their sweeping nature, I can only attempt to answer some of these challenges. Psychoanalytic insight helps us understand the crucial relationship between a social bond that recognizes the dignity of each one of us and the chance to develop even the most basic attributes of personality popularly associated with the self. Feminist theorists have notably criticized the overly rationalistic conception of the subject often presented in moral theory. Psychoanalytic insight can help us maintain what is right in that critique without arguing altogether against the place of reason in the exercise of freedom and moral responsibility in our lives. It can do so by providing us with an account of how our life as actively desiring subjects is not naturally given to human beings. Although we can all, *in principle*, claim our own desire, we can also be cut off from it if certain psychic laws and rules of separation are not incorporated into the basic social bonds of society and culture as ethical requirements that must be respected.

6. Again, I refer to the title of the book, *The Words to Say It*, and to Marie Cardinal, whose story I retell in this chapter.

7. But so far, this defense could be understood as only the need for psychoanalysis to the extent that it helps us to portray family dramas. Trinh T. Minh-ha, however, is clearly not limiting her account of intergenerational stories to individual families. She is, instead, retelling "Grandma's Story" to record the place of the storyteller as a feminist—understood as one who "partakes in the setting into motion of forces that lie dormant in us [because,] as African storytellers sing, 'the tongue that falsifies the word/taint of [her/]him that lies.' Because she who bears it in her belly cannot cut herself off from herself. Off from the bond of coming and going. Off from her great mothers." (See Minh-ha, *Woman, Native, Other*, p. 148.) Yet the oral traditions of wise women evoked by Minh-ha—African, Native American, South Asian—are not necessarily cultures that have embraced psychoanalysis. Indeed, the oral tradition of which Minh-ha writes serves a similar purpose by moving and shaking dormant, unconscious forces that drive us, and yet it does so within a different framework that may displace the one offered by psychoanalysis. So more needs to be said about the question I have posed, but first let me proceed by posing some further questions drawn from my original position. It is important, though, to note here that different psychoanalytic schools diverge not only on how we are constituted in our relationships with actual others but also on the question of whether there are others who are not analyzable as object relations between real people.

8. A full discussion of the distinction Lacan makes between drive and desire is beyond the scope of this chapter. For an excellent discussion of this distinction, see Charles Shepherdson, "The Epoch of the Body: On the Domain of Psychoanalysis," in *Perspectives on Embodiment*, eds. Gail Weiss and Honi Haber (Routledge 1999), pp. 183–211. For those interested in Lacan's unique reinterpretation of some of Freud's key ideas, Shepherdson's explanation of the drive, and particularly its relationship to the body may be helpful. Here I quote from an earlier, unpublished version of his essay:

The drive is not the residue of an archaic biological force that refuses to integrate itself into the symbolic norms of culture, it is rather an effect of the symbolic law that is nevertheless irreducible to representation. The drive can thus be understood as the bodily consequence of this partial character of representation. We may speak of this limit as a "failure" of representation, not simply in the abstract sense of a loss of meaning, or even in the sense of an excess of meaning (the fact that there is no closure to the movement of signification), but in the concrete sense that the drive will attempt to compensate for this failure (a lack in the Other) by producing something at the level of bodily *jouissance*—a *jouissance* that the subject does not desire. This is what Freud calls the "satisfaction of the drive," insofar as the drive is bound to suffering, or to corporeal satisfaction of the symptom, in which the desire of the desire is compromised. In contrast to desire, therefore, which is defined as never satisfied, the drive is a non-biological force (*Trieb, pulsion*) that always demands satisfaction (52, unpublished manuscript on file with the author).

9. It needs to be said here that in German idealism, and more specifically in Kant's critical idealism, the inclusion of the power to make ends is one that is stipulated in an ideal of the person, and not attributed to actual people. It is a presupposition, not an actuality taken for granted. This is a difference that makes a difference in my own feminism because part of my argument is that we must examine more carefully what is ethically entailed by this stipulation. It is not my position that we should not stipulate this power in the ideal of the person. Rather, we should understand that the position of the actively desiring subject is itself an ideal and one that feminism has explicitly demanded we give attention. Most schools of utilitarianism do attribute this power to people as a matter of fact. The fundamental difference here is that, if the power of actively desiring is a matter of fact, then it makes sense simply to speak of preferences as expressing those actual desires. However, if the power of desire is itself an ideal that people can in principle enact, then it makes sense to discuss how this position is to be represented, and once represented, respected. In chapter 3, I will return to why psychoanalysis insists on the space to represent desire rather than on the mechanisms for valuing preferences. For a classic statement of the stipulation of the free person in Kantian constructivism, which includes the power to actively set ends, see John Rawls, "A Kantian Conception of Equality," *Collected Papers* (Cambridge: Harvard University Press, 1999), pp. 254–266.

10. For an excellent discussion of how the dominant vision of the moral subject as man continues to influence moral philosophy, see Margaret Urban Walker, "Moral Understandings: Alternative 'Epistemology' For a Feminist Ethics," *Justice and Care: Essential Readings in Feminist Ethics*, ed. Virginia Held (Boulder, CO: Westview Press, 1995), pp. 139–152.

11. In the next two chapters I will discuss the significance in feminist theory of Cornelius Castoriadis's concept of the social imaginary. See generally Cornelius Castoriadis, *The Imaginary Institution of Society*, trans. Kathleen Blamey (Cambridge, MA: MIT Press, 1987).

12. This is not to say that the imaginary as well as the radical and productive imagination do not entail each other. Indeed, my argument is that they do. The imaginary and the imagination cannot be neatly distinguished; there is no one symbolic marker that designates the boundary between them. Rather, the

boundary is itself always reimagined and resymbolized in the process of ethical and political struggle within ourselves and with the world around us.

13. See Graciela Abelin-Sas, "The Internal Interlocutor" (unpublished essay on file with the author).

14. The need to maintain the ambivalence of the psychoanalytic idea of the imaginary explains why I insist on the distinction between the imaginary and the radical imagination. My distinction between the imaginary and the imagination differs from that of both Cornelius Castoriadis and Jacques Lacan, at least under one well-received notion of the Lacanian imaginary as the structural moment of the "mirror stage," or as the "phantasm" of the repressed pre-Oedipal phallic mother. For one discussion of Lacan's use of the imaginary, see Jacques Lacan, *The Four Fundamental Concepts: Book XI of the Seminar of Jacques Lacan* (New York: Norton, 1998). See also the discussion of the imaginary in Elizabeth Grosz, *Jacques Lacan: A Feminist Introduction* (New York: Routledge, 1990). Cornelius Castoriadis redefines the imaginary as the "radical imagination." I part ways with Castoriadis because I believe we need to maintain the distinction between the psychoanalytically informed concept of the imaginary and the radical or productive imagination. In his later work, Castoriadis explicitly anthropologized the imaginary. I disagree with this move in that it paradoxically puts blinders on the way in which we can endlessly reimagine the relationship between the imaginary and the symbolic and undercuts Castoriadis's own relentless critique of deterministic theories. See, for example, Cornelius Castoriadis's essay "The Imaginary" in *The World in Fragments: Writing on Politics, Society, Psychanalysis, and the Imagination* (Stanford: Stanford University Press, 1997), pp. 3–18. But I agree with Castoriadis in that there is never a neatly designated barrier between the imaginary and the radical imagination. Thus, like Castoriadis, I have opened up the Lacanian concept of the imaginary beyond its conceptualization in Lacan's own work.

15. See Carol S. Robb, *Equal Value: An Ethical Approach to Economics and Sex* (Boston: Beacon Press, 1995). Note, however, that the author uses the word *sex*, not *gender*, even though she points to the various forms of discrimination that keep women in positions of inequality. This is due to her sensitivity to the way in which discrimination law has been unable to reach the specific discrimination endured by lesbians (see pp. 91–111). For my discussion of how gender as a legal category has been detrimental to the claims of gays, lesbians, and transsexuals as well as to women of color generally, see *At The Heart of Freedom: Feminism, Sex, and Equality* (Princeton: Princeton University Press, 1998), pp. 3–8.

16. See generally Judith Butler, *Gender Trouble: Feminism and the Subversion of Identity* (New York: Routledge, 1990).

17. Special responsibility for our lives is one of Ronald Dworkin's two principles of ethical humanism. At the end of this chapter, I will return to discuss Dworkin's principle as it is relevant to the justice/care debate.

18. Butler's point is also that there is not something out there, whether we call it an essence or a symbolic order, that simply takes over and stamps us with a gender identity. Butler usually uses the word *gender*, but I believe that her emphasis on what I am calling the subjective aspect of the complex assumption of sexual identity is more consistent with the broader question of what remains "true" of sexual difference once we take into account some of the most fundamental in-

sights of psychoanalysis. Butler is concerned not only with what are popularly known as gender roles, but more broadly with the meaning of sexual differentiation, the ethical laws of kinship structures, and sexual desire. See Judith Butler, "Introduction," in *Bodies That Matter: On the Discursive Limits of Sex* (New York: Routledge, 1993) and *The Psychic Life of Power: Theories in Subjection* (Stanford: Stanford University Press, 1997). For an exchange between Butler and myself concerning the relationship between sexual difference and gender, see "The Future of Sexual Difference: An Interview with Judith Butler and Drucilla Cornell," *Diacritics* (Spring 1998): pp. 19–42.

19. For a thoughtful discussion of what is philosophically at stake in the debate, see Virginia Held, "Feminist Interpretations of Social and Political Thought," *Is Feminist Philosophy Philosophy?*, ed. Emanuela Bianchi (Chicago: Northwestern University Press, 1999), pp. 89–99. For an excellent survey of the different feminist perspectives on the debate, see *Justice and Care: Essential Readings in Feminist Ethics*, ed. Virginia Held (Boulder, CO: Westview Press, 1995).

20. Individuation describes a process through which we claim a personality that we call our own. Described as a process, individuation assumes that we are shaped but not determined by our primordial relations to others. Individualism is often popularly associated with the notion that human beings are born as individuals with given tastes, preferences, and other personality traits. Individualism is also associated with the ideology of capitalism in which we value ourselves above all others. Individuation is a process through which we claim as our own our ethical positions and values. There is nothing in the notion of individuation that describes an inherent connection between individuality and the placement of greater value on individual interests rather than familial or other group affiliations.

21. See Ruddick, "Injustice in Families: Assault and Domination," *Justice and Care: Essential Readings in Feminist Ethics*, pp. 215–217.

22. Marie Cardinal, *The Words to Say It*, trans. Pat Goodheart, preface and afterword by Bruno Bettelheim (Cambridge: VanVactor and Goodheart, 1983), pp. 5–6.

23. Cardinal, *The Words to Say It*, p. 4.

24. Cardinal, *The Words to Say It*, pp. 8–9.

25. Cardinal, *The Words to Say It*, pp. 9–10.

26. Cardinal, *The Words to Say It*, p. 13.

27. Cardinal, *The Words to Say It*, p. 91.

28. Cardinal, *The Words to Say It*, pp. 60–61.

29. Cardinal, *The Words to Say It*, p. 198.

30. Cardinal, *The Words to Say It*, p. 175.

31. Cardinal, *The Words to Say It*, p. 136.

32. Cardinal, *The Words to Say It*, pp. 139–40.

33. Cardinal, *The Words to Say It*, p. 140.

34. Cardinal, *The Words to Say It*, p. 88. But of course the Algeria being killed off was the mother's fantasy that was necessary for her status as a civilizing colonial. Here, again, we find Cardinal seeing through her mother's eyes. For the rise of the movement of national independence, which was a birth for those Algerians who participated in the struggle, was a death for the colonials who could not bear confronting the dying colonialism.

35. Cardinal, *The Words to Say It*, p. 290.

36. Cardinal, *The Words to Say It*, p. 293.

37. Cardinal, *The Words to Say It*, p. 60.

38. Cardinal, *The Words to Say It*, p. 299.

39. Cardinal, *The Words to Say It*, p. 68.

40. Cardinal, "Preface," *The Words to Say It*, p. xii.

41. Cardinal, *The Words to Say It*, pp. 259–260.

42. Cardinal, *The Words to Say It*, p. 264.

43. "Circus revolver" is one of Cardinal's metaphors for her mother's self-destructive behavior. See *The Words To Say It*, p. 293.

44. It is beyond the scope of this chapter to enter fully into the debate of whether and how Lacan changed his ideas concerning the relationship among the Symbolic, the Real, and the Imaginary over the course of his theoretical work. I am, however, inclined to agree with Derrida when he writes:

> Instead of trying to represent Lacan in general or psychoanalysis in general, I tried a long time ago to enter into a dialogue with a certain stage in Lacan's pedagogical and theoretical development. This was a stage represented by a number of important texts, such as the "Rome Discourse" or the "Seminar on The Purloined Letter." Looking at this stage, and this stage alone, I would agree with everything Drucilla said concerning Lacan's determination of the symbolic in terms of lack, the lack of the phallus, the role of the phallic mother here, and so on. I won't repeat what Drucilla carefully reviewed for us before.
>
> I would simply say that when looked at from the point of view of that stage, Lacan was *too* philosophical—too much of a philosopher. In his own original way, he was repeating a very old and very powerful philosophical discourse. At least that's what I tried to show. Later on, in the seminars that followed, he made other statements that complicated a number of things. For example, he finally dropped, or gave up, the triangle of the Symbolic, the Real, and the Imaginary. And so everything became more complicated.

(See Jacques Derrida, "Opening Remarks," in *Is Feminist Philosophy Philosophy?*," ed. Emanuela Bianchi [Evanston, IL: Northwestern University Press, 1999], pp. 13–14. For my own discussion of this stage in Lacan's work, see Drucilla Cornell, "Opening Remarks," pp. 5–7, in the same volume.)

45. For an excellent discussion of Lacan's concept of the Symbolic, see Charles Shepherdson, *Vital Signs: Nature, Culture, Psychoanalysis* (New York: Routledge, 2000), pp. 37–39, 62–69.

46. *Stiffed* is the title of Susan Faludi's influential book on masculinity, particularly white working class masculinity in U.S. society. As Faludi rightfully notes, many men feel the problem is feminism. The feeling—if we can just get rid of feminism, we can go back to the good old days when there were real fathers—is reflected in the father's movement. For my own analysis of the father's movement, see "What and How Maketh a Father?: Equality Versus Conscription," *At the Heart of Freedom*, pp. 131–150.

47. See Simone de Beauvoir, *The Second Sex*, trans. H. M. Parshley (New York: Vintage Books, 1974).

48. For a brilliant and detailed reading of what Lacan means by this statement, see Paul Verhaeghe, *Does the Woman Exist?: From Freud's Hysteria to Lacan's Feminine*, trans. Marc Du Ry (New York: Other Press, 1998).

49. See Joan Riviere, "Womanliness as a Masquerade," in *Formations of Fantasy*, eds. Victor Burgin, James Donald, and Cora Kaplan (London: Methuen, 1986).

50. For example, Nicola Lacey has questioned my own use of Lacan because she reads him this way. See Nicola Lacey, *Unspeakable Subjects: Feminist Essays in Legal and Social Theory* (Oxford: Hart Publishing, 1998), pp. 205–207.

51. See Judith Feher Gurewich, "Is the Prohibition of Incest A Law?" (unpublished manuscript on file with the author), p. 5.

52. See Judith Feher Gurewich, "The *Jouissance* of the Other and the Prohibition of Incest: A Lacanian Perspective" (unpublished manuscript on file with the author), p. 3.

53. Gurewich, "The *Jouissance* of the Other," p. 5.

54. Gurewich, "The *Jouissance* of the Other," p. 8.

55. Gurewich, "The *Jouissance* of the Other," p. 11.

56. For an excellent argument that it is the desiring mother and thus the mother of the symbolic who plays a central role in the child's crisis, see Charles Shepherdson, *Vital Signs*, pp. 60–72.

57. See Paul Verhaeghe, "The Collapse of the Function of the Father and Its Effect on Gender Roles" (unpublished manuscript on file with the author), p. 12.

58. For excellent modern discussions of Kantian moral philosophy, see Christine M. Korsgaard, *Creating the Kingdom of Ends* (Cambridge: Cambridge University Press, 1996).

59. For excellent discussions of Kant's moral philosophy, see the debate in *Justice and Care*. See also Louise M. Antony, *A Mind of My Own: Feminist Essays on Reason and Objectivity* (Boulder, CO: Westview Press, 1993). For feminist defenses of Kant and modern Kantian moral theory in the form of social contract theory, see Jean Hampton, "Feminist Contradaranism," *A Mind of My Own: Feminism Essays on Reason and Objectivity* (Boulder: Westview Press, 1993), pp. 227–257.

60. For an excellent discussion of how the cut from one's own desire is pervasive in patients who suffer from severe depression, see Darlene Bregman Ehrenberg, *The Intimate Edge: Extending the Reach of Psychoanalytic Interaction* (New York: W.W. Norton and Company, 1992).

61. See Castoriadis, *The Imaginary Institution of Society*, p. 107.

62. As I have argued elsewhere, I do not think such a neat philosophical explanation of the role of the Real survives its deconstruction by Jacques Derrida. See Drucilla Cornell, "Opening Remarks," in *Is Feminist Philosophy Philosophy?*, ed. Emanuela Bianchi (Evanston, IL: Northwestern University Press, 1999), pp. 3–9; see also Drucilla Cornell, "Rethinking the Beyond of the Real," *Levinas and Lacan: The Missed Encounter*, ed. Sarah Harasym (Albany: State University of New York Press, 1998), pp. 139–181.

63. See Castoriadis, *The Imaginary Institution of Society*, p. 142.

64. My own interpretation of Castoriadis is that the fundamental phantasy of which he writes is close to Kant's notion of the schemata, associated in Kant with the transcendental imagination. While this imagination and these schemata are the backdrop of everything we can know, they can be grasped by us only through a transcendental deduction. Kant wisely differentiates the schema, which we must represent as already in place whenever we begin to seek knowledge, from any notion of the productive imagination. (See Immanuel Kant, *Critique of Pure Reason*, trans. Norman Kemp Smith [New York:

St. Martin's Press, 1969] and *The Critique of Judgement*, trans. James Creed Meredith [Oxford: Oxford University Press, 1952].) Castoriadis, on the other hand, roots the structural predominance of the imaginary in an ontological rather than deontological account of the imagination. (See *The Imaginary Institution of Society*, pp. 146–160.) I agree with Castoriadis that the imaginary and the symbolic cannot and should not be neatly distinguished, and that the imaginary is always in play in the struggle for our individual and collective freedom. However, I would not defend his claim that, in the relationship between the imaginary and the symbolic, the imaginary is necessarily predominant. My disagreement with Castoriadis stems from our diverging views on the enhancing power of language. See Toni Morrison's differentiation between living language and dead language, in *The Nobel Lecture in Literature* (New York: Alfred A. Knopf, 2000), pp. 13–14. Over and over again, Castoriadis refers to language, any language, as a barrier that the imaginary breaks through. For me, the symbolic and the imaginary are co-relative. If the imaginary and the symbolic feed on each other, it would be difficult, if not impossible, to fix the distinction between them in time or in causal power. In the name of Castoriadis's important project, which is to give due respect to the creativity of the imaginary, I would argue that the defense of the predominance of the imaginary actually undercuts his own theoretical aspiration. For a longer discussion of Castoriadis, see Drucilla Cornell, *Who Is Afraid of Disorderly Conduct?* (forthcoming).

65. See generally Martin Heidegger, *Kant and the Problem of Metaphysics* (Bloomington: Indiana University Press, 1997).

66. See Drucilla Cornell, "Spanish Language Rights: Identification, Freedom, and the Imaginary Domain," *Just Cause: Freedom, Identity and Rights* (Lanham, MD: Rowman & Littlefield, 2000), pp. 129–153.

67. See Castoriadis, *The Imaginary Institution of Society*, p. 107.

68. See Jessica Benjamin, *The Bonds of Love: Psychoanalysis, Feminism and the Problem of Domination* (New York: Pantheon Books, 1988).

69. See Cornell, "Preface," *At the Heart of Freedom*, pp. ix-xii.

70. See Jessica Benjamin, *The Bonds of Love: Psychoanalysis, Feminism and the Problem of Domination* (New York: Pantheon Books, 1988).

71. It is important to note that such ideals are not identical with the Lacanian subject. The subject for Lacan is the symbolic inscription of the human creature into a symbolic order that both marks the body as sexed and at the same time produces neurotic structures that can be traced back to the patient's own inheritance, or lack thereof, of the paternal function. The subject is also born through an inevitable moment of separation, which I have argued can be understood as both the dignity of the children and that of their primary others. But these two versions of the psychoanalytic subject share this in common: the psychoanalytic subject is formed through its relation to the Big Other and the fantasies it has produced to protect the child against his or her own incestuous longing (again, incest is used in Gurewich's reading of Lacan as a metaphor for the longing to dissolve one's separate existence into that of the primary Other). In this sense, as I have argued earlier, the subject and the subject of the unconscious are tied together. Lacan's statements about the dangers of strengthening the ego refer to his concern for the defenses maintained by the ego against his or her own unconscious and how these can further bury the subject. But all the same, the ego,

or what Anglo-American analysts call the *self*, is still a structural moment. We never simply overcome the ego, or what Lacan calls the Imaginary. Thus I have argued that it is perfectly consistent with Lacan's understanding of the imaginary, which is also the place of the ego, to defend an ideal of the self, similar to the one advocated by Benjamin. However, I would now argue that we need a more expansive definition of the imaginary than the one offered by Lacan.

72. Teresa Brennan has also argued that feminism can push us toward changing rigid gender roles. See Teresa Brennan, *The Interpretation of the Flesh* (New York: Routledge, 1993).

73. See Jessica Benjamin, *Shadow of the Other: Intersubjectivity and Gender in Psychoanalysis* (New York: Routledge, 1998), p. xvii (citations omitted).

74. See Drucilla Cornell, *The Imaginary Domain: Abortion, Pornography, and Sexual Harassment* (New York: Routledge, 1995).

75. See John Rawls, "The Idea of an Overlapping Consensus," in *Political Liberalism* (New York: Columbia Press, 1993), pp. 134–172.

76. See Cornell, *At the Heart of Freedom*, pp. 151–173.

77. See bell hooks, *All about Love: new visions* (New York: William Morrow, 2000).

78. See Elisabeth Bronfen, *Over Her Dead Body: Death, Femininity, and the Aesthetic* (Manchester: Manchester University Press, 1992), pp. 34–35. In addition, Elisabeth Bronfen and I differ from Lacan's reading of Heidegger.

79. Charles Shepherdson discusses why the mother is not simply imaginary. See *Vital Signs*, pp. 60–61.

80. As Gurewich explains, in contradiction to some of the traditional interpretations of Lacan:

> Yet the experience of the child as the apple of the mother's eye, as the exclusive object of the mother's desire, of course presupposes that the mother is a desiring being, in other words, that she wants something that she does not have. The experience of being the object of the Other's desire of course implies that the subject registers that he could also fail to occupy that position. In Lacanian terms, this translates as: the child must come to grips with the fact that the mother is lacking, and that something or someone is able to fill that lack. This is a crucial aspect of Lacan's understanding of the oedipal dynamics. For the child to discover what he is for the other requires that he also.
>
> This is why Lacan says that castration is the ability to recognize the lack in the (m)Other. This does not mean, of course, that the mother lacks a penis, but rather, as Freud noted in his last essay on femininity, that she lacks the phallus, that she lacks that which could bring her fulfillment. If the mother's desire cannot view her child as a separate being whom she can admire, love, and desire, the child will instead encounter the mother's *jouissance*, that is, a realm of enjoyment that is not symbolized, something akin to Melanie Klein's definition of the maternal superego or Kohut's selfobject (4–5, unpublished manuscript on file with the author).

See also Judith Feher Gurewich, "The Subversive Value of Symbolic Castration: The Case of Desdemona." *JPCS: Journal for the Psychoanalysis of Culture and Society* 2 (Fall 1997): pp. 61–66. In that sense, I would argue, the question of the Real of sexual difference is not directly addressed by Lacan. The question that Freud attempted to answer by reverting to biological determinism seems to fall out of Lacan's consideration; in the final consideration he does not tell us what becomes of a woman who has resolved her Oedipal complex. I'm certainly not forgetting Lacan's seminar *Encore* in which he suggests that women's *jouissance* is not all inscribed in the phallic function. Nevertheless, isn't this earth shattering revelation more related to men's fantasy of women than to women themselves?

NOTES TO CHAPTER 3

1. See Trinh T. Minh-ha, *Woman, Native, Other: Writing Postcoloniality and Feminism* (Bloomington: Indiana University Press, 1989), p. 122.
2. Minh-ha, *Women, Native, Other*, p. 123.
3. See Gayatri Chakravorty Spivak, *A Critique of Postcolonial Reason: Toward a History of the Vanishing Present* (Cambridge, MA: Harvard University Press, 1999), pp. 198–311.
4. See Gayatri Chakravorty Spivak "Subaltern Studies: Deconstructing Historiography," *In Other Worlds* (New York and London: Methuen, 1987), pp. 197–221.
5. See Spivak, *A Critique of Postcolonial Reason*, pp. 4, 9.
6. Spivak, *A Critique of Postcolonial Reason*, p. xi.
7. In *A Critique of Postcolonial Reason*, Spivak describes psychoanalysis "as a technique for reading the pre-emergence (Raymond William's term) of narrative as ethical instantiation" (4).
8. How does a writer take notice of women's stories when they continually fall into invisibility and silence and when a concerted effort is undertaken to preserve the voices and their presence in history? As a writer, Spivak does so by relying on Jacques Lacan's term *foreclosure* in two ways. First, she relies on this term to show how what is missing in a literary text or historical narrative leaves its mark through the traces of its expulsion. Second, she uses the concept in a more technical sense: it is an energetic defense of the ego that drains the incompatible idea or object of affect and allows the ego to disavow aspects of the external world as having any reality or significance. For Spivak, this form of *foreclosure* serves the energetic defense of the civilizing mission in which those who carry it out can neither see nor feel the violence they commit in its name. The incompatibility of the civilizing mission, with many of the values that Western societies hail as their hallmark, is the incompatible object that demands that relationships with the natives be drained of the affects that human beings are supposed to show one another in a "civilized society." Among these values, of course, I would make central the dignity of the subject and the entitlement of each person to demand respect for her freedom. The disavowal can take at least two forms: 1) it completely denies that there is an incompatible object—for instance, the colonizer expunges the colonized from humanity so that ethical standards for how one interacts with other human beings no longer apply; 2) it disavows what is happening altogether and therefore does not record it or distorts it horribly.
9. See Spivak, *A Critique of Postcolonial Reason*, p. 292.
10. Spivak, *A Critique of Postcolonial Reason*, p. 244.
11. Spivak, *A Critique of Postcolonial Reason*, p. 245.
12. Spivak, *A Critique of Postcolonial Reason*, p. 287.

13. For Cornelius Castoriadis's account of the social imaginary, see *The Imaginary Institution of Society*, trans. Kathleen Blamey (Cambridge, MA: MIT Press, 1987), pp. 135–146.

14. See Spivak, *A Critique of Postcolonial Reason*, p. 232.

15. Spivak, *A Critique of Postcolonial Reason*, p. 233.

16. Spivak, *A Critique of Postcolonial Reason*, p. 231.

17. Spivak, *A Critique of Postcolonial Reason*, p. 233.

18. Spivak, *A Critique of Postcolonial Reason*, p. 237.

19. Spivak, *A Critique of Postcolonial Reason*, pp. 233–34.

20. See Dominick La Capra, *Rethinking Intellectual History: Texts, Contexts, Language* (Ithaca: Cornell University Press, 1983).

21. See Spivak, *A Critique of Postcolonial Reason*, p. 207.

22. Spivak, *A Critique of Postcolonial Reason*, p. 242.

23. Spivak, *A Critique of Postcolonial Reason*, p. 235.

24. Spivak, *A Critique of Postcolonial Reason*, p. 241.

25. Spivak, *A Critique of Postcolonial Reason*, p. 239. Indeed, her critique of other activist philosophers turns on the way they render themselves "transparent" when they purport to be speaking for the people themselves and when they represent groups in struggle who do not need any representation other than self-representation. As Spivak rightly points out, representation means both portrait and proxy. *Portrait* refers to the representational consciousness by which we see our world. *Proxy* means the political form of representation in which one person or persons represents the group or—as can still be the case in many local organizations—the group struggles to articulate its own interests, as it does away with the idea of a proxy. I will return to the practice of radical democracy in forms of cooperative organizing among workers that are developing throughout the world. Spivak herself focuses on the cooperative movement of textile workers. I will focus on recent efforts to organize nannies' collectives. Both of these examples have implications for how the working class is represented because workers, particularly women workers, are insisting that they have interests they wish to represent as those of women workers in organized cooperatives. The result is that, by representing themselves as such, they are no longer invisible as workers since they have claimed themselves as a part of the working class.

26. Richard Rorty brilliantly exposes the fallacies in the idea of a mirroring consciousness in *Philosophy and the Mirror of Nature* (Princeton: Princeton University Press, 1980).

27. See Spivak, *A Critique of Postcolonial Reason*, p. 208.

28. Anthony Cascardi provides an excellent critique of Jurgen Habermas's reading of Kant's *Critique of Judgment*. See Anothony Cascardi, *Consequence of Enlightenment* (New York: Cambridge University Press), pp. 132–179.

29. See Kant, *The Critique of Judgement*, trans. James Creed Meredith (Oxford: Oxford University Press, 1952), p. 29.

30. Although there is language in *The Critique of Judgment* that seems to justify this interpretation, it simply cannot be what Kant means. This "cure" for subjectivism would have effectively undermined his project of finding a place in his critical philosophy for the subject who feels and suffers. Yet this, precisely, is the mistake of Hannah Arendt, who, like Jurgen Habermas, collapses the *sensus communis aestheticus* into the *sensus communis logicus*. See Anthony J. Cascardi,

"Communication and Transformation: Aesthetics and Politics in Habermas and Arendt, *Consequences of Enlightenment*, pp. 132–74.

31. As Kant explains a judgment of taste:

> That is to say since the freedom of the imagination consists precisely in the fact that it schematizes without a concept, the judgment of taste must found upon a mere sensation of the mutually quickening activity of the imagination in its *freedom* and of the understanding with its *conformity of law*. It must therefore rest upon a feeling that allows the object to be estimated by the finality of the representation (by which one object is given) for the furtherance of the cognitive faculties in their free play. Taste, then, as a subjective power of judgment contains a principle of subsumption, not of intuitions under *concepts* but of the *faculty* of the intuitions or presentation i.e., of the imagination, under the *faculty* of concepts, i.e., the understanding, so far as the former *in its freedom* accords with the *latter in conformity with its law* (*The Critique of Judgement*, p.143).

All we can know of these judgments as such is their formal features. But one aspect of our pleasure and our pain is that we might be able to share these feelings with others. Indeed, Kant emphasizes that it is our sociality as human beings that promotes us to seek to communicate our feelings to others and that a man on a desert island would not care about the beautiful or the sublime because their would be no one with whom to share his feelings. This emphasis on the sociability of our interest in the beautiful in particular has undoubtedly influenced the reading of Kant that interprets common sense as something actually existing in a governable community such that we can depend on our ability to communicate our feelings and thus translate them into a judgment we share with others. But if this were the case, then Kant would lose the uniqueness of what he calls subjective universal validity. He would be collapsing our judgment into the objective world of shared conventions of an existing group.

32. Kant, *The Critique of Judgment*, p. 153.

33. I am of course aware that, in *The Critique of Judgment*, Kant sometimes makes comments that promote the idea of his reliance on elitist groups to support aesthetic judgments. This becomes evident when he connects the ability to feel nature as a sublime object with a high degree of civilization. However, this is not at the heart of his argument. It only reflects his prejudice regarding who is and is not civilized.

34. Kant, *The Critique of Judgment*, p. 153.

35. Kant, *The Critique of Judgment*, p. 151.

36. Indeed, Kant connects the *should be* inherent in reflective judgment with the necessary postulation of an *a priori* principle, which in turn relates to the general problem that plagues him—how synthetic, as opposed to analytic, judgments are possible at all. The attempt to bring closure by defining the community as one in the present such that convention, rather than sense, is the basis of judgment has prompted a conventionalist reading of Kant. The realization that Kant is appealing to the collective reason of mankind has alternatively led to the interpretation that these idealized representatives actually give us rationally generated standards of taste. Yet there are theorists who have adequately grappled with the infinitely open process of reflective judgment. Think, for example, of John Rawls and his famous aesthetic idea of the "veil of

the ignorance." In his later work, Rawls clearly defends the veil of ignorance as a representational device, although he never explicitly defends the veil of ignorance as an aesthetic idea. In *A Theory of Justice*, Rawls defends at times the veil of ignorance as if it were an *aesthetic idea*, the *adequate representation of the imagination* that could represent what rational individuals could be reasonably expected to agree upon. In this sense, the experiment in the hypothetical imagination could generate principles of justice. However, once one understands that an aesthetic idea cannot directly represent the concept in reflective judgment, one can also deduce that the veil of ignorance cannot claim to be the sole and only adequate representation, which is what, in some of his writing, Rawls seems to claim. Aesthetic ideas are still extremely important in helping us see what justice demands of us ethically. Rawls's aesthetic idea of the veil of ignorance changed the direction of political philosophy. Nevertheless, an aesthetic idea should not bring closure to the human imagination. There can always be other figures. These aesthetic ideas cannot directly generate principles of justice. See John Rawls's discussion of the veil of ignorance in *A Theory of Justice* (Cambridge, MA: Harvard University Press, 1999). His conception of the veil of ignorance in *A Theory of Justice* differs from his other conception of in his later work, *Political Liberalism*. See John Rawls, *Political Liberalism* (New York: Columbia University Press, 1996).

37. For a definition of what Kant means by "public sense," see *The Critique of Judgement*, p. 153.
38. Kant, *The Critique of Judgement*, p. 151.
39. Kant, *The Critique of Judgement*, p. 82.
40. Kant, *The Critique of Judgement*, pp. 175–176.
41. Kant, *The Critique of Judgement*, pp. 111–112.
42. In the example he gives to show that his idea of the sublime does not turn us to any specific object realm, Kant shows his crass Eurocentrism and his masculine bias: "For what is it that, even to the savage, is the object of the greatest admiration? It is the man who is undaunted, who knows no fear, and who, therefore, does not give way to danger, but sets manfully to work with full deliberation. Even where civilization has reached a high pitch there remains this special reverence for the soldier" (112). His Eurocentrism and masculine bias are subtler in his idea of the sublime than in his analysis of the operation of the beautiful. Although it demonstrates the limits of Kant's own moral imagination, the example makes evident that the feeling of the sublime that evokes our freedom is not necessarily best imagined in our relationship to natural objects.
43. See Friedrich Schiller, "On the Sublime," *Friedrich Schiller: Essays*, eds. Walter Hinderer and Daniel O. Dahlstrom (New York: Continuum Publishing Company, 1998), p. 35.
44. Schiller is very clear that the aesthetic should never be entrapped by the moral:

> In making aesthetic judgments we veer from actual things to possibilities and rise up from the individual to the species, while in making moral judgments we descend from the possible to the actual and enclose the species within the limitations of the individual. No wonder, then, that we feel ourselves broadened in aesthetic judgments, but confined and restricted in moral judgments.
>
> From all this it follows, then, that moral and aesthetic judgments, far from supporting one another, stand rather in each other's way, since

they point the mind in two diametrically opposed directions. For the law abidingness demanded by reason as a moral judge is inconsistent with the lack of restraint required by imagination as an aesthetic judge. Thus an object will be all the less suitable to some aesthetic use the more it qualifies for some moral use.

(See Friedrich Schiller, "On the Pathetic," *Friedrich Schiller: Essays*, eds. Walter Hinderer and Daniel O. Dahlstrom [New York: Continuum Publishing Company, 1998], pp. 64–65.)

45. Can this form of witnessing make a claim to universalality? The universality of Kant's *sensus communis aestheticus* is peculiar because it never exists as such but is still evoked as an *ought to be* that makes aesthetic or reflective judgment possible. The presumption that all others included in the reach of humanity should agree with us in our judgment is evoked as the only basis of and support for the possibility of reflective judgment. But it is a mistake to argue that this peculiar form of universality is simply too weak to play any significant role in transnational feminism. Terry Eagleton has made this mistake by underestimating the power of Kant's aesthetic community. (See Terry Eagleton. "The Kantian Imaginary," *The Ideology of the Aesthetic* (Oxford: Blackwell Press, 1990), pp. 7–101.) The experience of witnessing to the sublimity of the struggles of the gendered subaltern has an intensity all its own. The intense feeling, which the sublime calls forth, allows us to come together at the moment of witnessing; this intensity of feeling is irreducible to our conscious motivations, which lead us to develop strategic alliances in our day-to-day struggle. The intertwining of dignity with the aesthetic quality of the sublime, however, introduces an ethical moment into reflective judgment that reinforces dignity even if it does not, in *that form of judgment*, seek to offer a full philosophical defense of dignity. Yet without dignity in the picture, the emotional response to the presence of a sublime object before us will not be aroused. To view the struggles of the subaltern as sublime demands that white Anglo women expand their mentality in Kant's sense. All too often in recent human rights discourse, we see the gendered subaltern represented as victims to whom we must offer aid.

46. See Spivak, *A Critique of Postcolonial Reason*, p. 310.

47. Spivak, *A Critique of Postcolonial Reason*, p. 282.

48. Since Western philosophy's very inception, witnessing as a kind of vision or beholding has been inextricably linked to care. Perhaps the most densely philosophical statement of this can be found in Martin Heidegger's immortal work, *Being and Time*. "The care [*Sorge*] for seeing," Heidegger says, " is essential to man's Being. . . . Being is that which shows itself in the pure perception which belongs to beholding, and only by such seeing does Being get discovered. Primordial and genuine truth lies in pure beholding." (See Martin Heidegger, *Being and Time* [New York: Harper & Row, 1962], p. 215.) For a lucid elaboration of this point from a unique theoretical perspective that brings together psychoanalysis, Platonic philosophy, and continental thought, see Kaja Silverman, *World Spectators* (Stanford: Stanford University Press, 2000).

49. See Schiller, "On the Sublime," *Friedrich Schiller: Essays*, eds. Walter Hinderer and Daniel O. Dahlstrom (New York: Continuum Publishing Company, 1998), p. 42.

50. See Spivak, *A Critique of Postcolonial Reason*, p. 235.

51. See Schiller, "On the Pathetic," p. 66.

52. See Spivak, *A Critique of Postcolonial Reason*, p. 311.
53. Spivak, *A Critique of Postcolonial Reason*, pp. 294–295.
54. See Hannah Arendt, *Lectures on Kant's Political Philosophy*, ed. Ronald Beiner (Chicago: University of Chicago Press, 1982), p. 77.
55. Without the attribution of dignity there is no feeling of sublimity because there is no representation of moral resistance. When we witness to the sublimity of the event, we are refusing to evaluate them in terms familiar to us. Kelly Oliver has written of my own conception of feminist witnessing:

> With her notion of feminist witnessing, however, Cornell imagines that rather than entitlement and solidarity following from a group formation based on recognition of others like me or us—the equation of difference—entitlement and solidarity follow from admitting that we are not alike, that we cannot be alike, that in important ways we cannot be alike. With her notion of feminist witnessing, she suggests an alternative to legal recognition. The type of recognition that is part of witnessing is the recognition of the impossibility of recognition, recognition as a confession or admission that we are not in a position to validate the other's experience. As Cornell maintains, this witnessing begins in the recognition of an ethical call from the other. Rather than the process of recognition starting from the rational autonomous subject of legal authority, the process of recognition begins with a call from the other. Yet, the call can only be answered, or perhaps even heard, if we give up the position of the master-authorized-to-recognize and validate the other.

(See Kelly Oliver, "Recognition, Witnessing, and Identity: Drucilla Cornell on Family Law," *Subjectivity without Subjects: Abject Fathers to Desiring Mothers* [Lanham, MD: Rowman & Littlefield, 1998], p. 93.) I agree with much of what Oliver writes, but I disagree with her when she insists that recognition is the impossibility of recognition, and that it is to this impossibility that we must confess.
56. Ibid., p. 93.
57. See Gayatri Chakravorty Spivak, "Philosophy," *A Critique of Postcolonial Reason: Toward a History of the Vanishing Present*, pp. 37–67.

NOTES CHAPTER 4

1. See Chandra Talpade Mohanty, "Crafting Feminist Genealogies: On the Geography and Politics of Home, Nation, and Community," *Talking Visions: Multicultural Feminism in a Transnational Age*, ed. Ella Shohat (Cambridge, MA: MIT Press, 1998), p. 486. The phrase was coined by Angela Davis.
2. See interview with Zoila Rodriguez, chapter 5, p. 132.
3. Jacqueline Martinez uses the semiotics of Charles Sanders Pierce to help clarify the necessarily communicative and discursive aspects of identification, particularly racial and ethnic identifications. "To restate the case," she argues, "notions of ethnic or racial heritage are representations (enabled by interpretants) that also become self representations (thus creating new interpretants that maintain or change the content of the representations)." (Jacqueline Martinez, *Phenomenology of Chicana Experience and Identity: Communication and Transformation in Praxis* [Lanham, MD: Rowman & Littlefield, 2000], p. 17.) Allow me to interrupt Martinez for a moment and explain what an interpretant is for readers unfamiliar

with the work of Pierce. An interpretant is a frame of reference we take for granted that enables us to give meaning to other signs. Interpretants always have a futural dimension, which is why the semiotic field is infinitely open. Sometimes we take interpretants for granted because they become conflated with representations we simply take as the way of the world. Habits of mind and dispositions toward the world are how interpretants come to be hidden. However, since interpretants are necessarily caught up in a chain of other interpretants giving meaning and significance to our world, they can always be put into question. The mood of interpretants is the "would be" as we allow a particular representation guide us in our actions. We act through our interpretations as we take a particular representation and let it guide us in our relationship to the world and to that of other interpretants. (See Charles Sanders Pierce, "Prolegomena to an Apology for Pragmatism," *Pierce on Signs*, ed. John Hoopes [Chapel Hill: University of North Carolina Press, 1991], pp. 249–252. For my own discussion of the future orientation of interpretants and why our interpretations of the world proceed through recollective imagination, see *Transformations: Recollective Imagination and Sexual Difference* [London: Routledge, 1993], pp. 26–29.) I want now to return to Martinez because it is she who brings Pierce's insights to bear particularly on racial and ethnic identification and who shows why the idea of self-representation needs neither "individualism" nor the "philosophy of consciousness," as Jürgen Habermas calls it. As she suggests:

> I come into the world of ready-made ethnic and racial representations that do or do not become self-representations based on notions of my own and others' nature—I either am or am not black; Spanish either is or is not my first or predominant language; I can either take the security of my housing, food, and clothing situation for granted or I cannot; and so on. The moment I recognize myself (preconsciously) as on one side or another of those representations, they become self-representations. But in making them self-representations, they do not simply replicate the given representations themselves. Rather, the representations themselves become producers of representations in those very moments in which I preconsciously take them up. This is an especially salient point when one considers the massive flow of racial and ethnic representations that suffuse virtually every kind of mass communication, the awesome materiality of discourse (to borrow a phrase from Foucault), and the way in which the person, situated as an active nexus point within this complex of communication and temporality, habitually takes up these ready-made representations and makes of them self-representations.

4. See Ernesto Laclau, "Identity and Hegemony," in Judith Butler, Ernesto Laclau, and Slavoj Žižek, *Contingency, Hegemony, Universality: Contemporary Dialogues on the Left* (London: Verso, 2000), pp. 40–89. On the one hand, I agree with Laclau that "there is an ethical *investment* in particular normative orders, but no normative order which is, in and for itself, ethical" (81). (For a more general discussion of this from the perspective of an ethics of deconstruction, see my book, *The Philosophy of the Limit*, [London: Routledge, 1992].) A society that is in and for itself ethical would be one in which no further democratic challenge would be possible. Those of us on the Left know it as Marx's dream of a fully transparent society in which representation and thus the state would no longer be necessary.

Without the false consciousness inherent in a pre-communist society in which every class presented its particularity as realized universality, human beings would finally be able to know their true needs and interests. We could finally live together under Marx's brave configuration of the principle of a truly communist society: "From each according to his ability; to each according to his needs." To quote Laclau: "Hegemony is, in this sense, the name for this unstable relation between the *ethical* and the *normative*, our way of addressing this infinite process of investments which draws its dignity from its very failure. . . . I would say that 'hegemony' is the theoretical approach which depends on the essentially ethical decision to accept, as the horizon of any possible intelligibility, the incommensurability between the ethical and the normative" (81). I disagree with him, however, when he argues further that "only that aspect of a decision which is not predetermined by an existing normative order is, properly speaking, ethical" (82). I would rather insist that there is no such thing as a purely ethical act in Laclau's sense. Yes, we can know that it is impossible for any society to claim the realization of the Good in the name of "good democracy." But what we are left with is a field of "competing universalities," to use Judith Butler's phrase. (See her essay of the same name in *Contingency, Hegemony, Universality*, pp. 136–138.)

We cannot simply step out of our own ethical investments and knowingly assert, in the spirit of Kierkegaard, that "the moment of the decision is the moment of madness." Nor can we follow Laclau and translate such an assertion to mean that, as a rigid matter of subjectivity, "this is the moment of the subject before subjectivation" (79). In other words, we cannot know ourselves to be acting "madly" except retrospectively and then only by describing it as madness, let alone as the true ethical act for which we are already providing a normative framework. Often—forgive me an aside here—the description takes the form of "Wow, what a man!" (Think of Slavoj Žižek's list of male heroes in his beloved films.) In this way, we are left to act and to name universals in order to enter the fray of struggle for hegemony. I thus agree with Butler when she writes, "But if various movements speak in the name of what is universally true for all humans, and not only do not agree on the substantive normative issue of what that good is, but also understand their relation to this postulated universal in semantically dissonant discourses, then it seems that one task for the contemporary intellectual is to find out how to navigate, with a critical notion of translation at hand, among these competing kinds of claims on universalization" (163). One aspect of this navigation is to understand that the names we give to universals have an aesthetic dimension in the sense of Kant's *Critique of Judgment* (and the way I defend his conception of reflective judgment in chapter 3).

5. See interview with Zonia Villanueva, chapter 5, p. 129.
6. John Rawls has argued that as a matter of universally respecting justice within liberal society, we must have a sense of responsibility to the nation-state in which we are citizens. According to Rawls, one of the principles of justice between free and democratic peoples is that "[p]eoples (as organized by their governments) are free and independent and their freedom and independence is to be represented by other peoples." (See John Rawls, "The Law of Peoples," *Collected Papers*, p. 541.) This cosmopolitan principle makes perhaps the most sense when considered in light of his early and very influential claim that "the state is [to be] viewed as the association of citizens to regulate their pursuit of their profoundest interests and their fulfillment of their most solemn obligations and

to give form to their relations in a manner acceptable to each from an original position of equal liberty." (See John Rawls, "Constitutional Liberty and the Concept of Justice," *Collected Papers*, p. 90.)

7. See Inderpal Grewal, "On the New Global Feminism and the Family of Nations," in *Talking Visions*, p. 516.

8. There is a rich, constantly expanding leftist literature addressing what, in early 1999, the editors of the *New Left Review* quite accurately dubbed "the imperialism of human rights." They were of course referring to the war in Serbia and the U.S. effort to "help" Kosovar Albanians. The thrust of the critique has as its larger object what Noam Chomsky understands to be "the guiding criteria for humanitarian concerns: atrocities for which we bear responsibility, and which we could easily mitigate or terminate, do not 'have anything to do with' us and should be of no particular concern; worse still, they are a diversion from the morally significant task of lamenting atrocities committed by official enemies that we can do little if anything about." (See Noam Chomsky, *A New Generation Draws the Line: Kosovo, East Timor, and the Standards of the West* [London: Verso, 2000], p. 81.) Hence the central thesis of Chomsky's most recent book on East Timor and Kosovo:

> Indonesia's non-existent claim to sovereignty in East Timor was accorded the most delicate respect under the operative principles of the enlightened states. They insisted that its military forces must be assigned responsibility for security while they were conducting another reign of terror. As for Kosovo, the US and its allies require that it must remain under Serbian sovereignty, probably out of fear of a "greater Albania." But the sovereignty that NATO insists upon in Serbia is "trumped" by its claim that it is defending human rights, unlike East Timor, where non-sovereignty "trumps" any concern for the human rights that the leaders of NATO are brutally violating (25).

Edward Said, longtime intellectual compatriot of and political collaborator with Chomsky, was true to this critical spirit when, in early 1999, he wrote apropos of Kosovo, "Nothing of what the US or NATO does now has anything really to do with protecting the Kosovars or bringing them independence: it is rather a display of military might whose long-range effect is disastrous, just as is a similar policy in the Middle East. . . . The humanitarian concerns expressed are the merest hypocrisy since what really counts is the expression of US power, and the rest of the world, including the UN, be damned." (See Edward Said, "Protecting the Kosovars?," *New Left Review*, March/April 1999, pp. 74–75.) Former member of the Yugoslav army and perhaps the most prolific intellectual of the past decade, Slavoj Žižek, agreed wholeheartedly:

> . . . we see NATO, the armed hand of the new capitalist global order, defending the strategic interests of capital in the guise of a disgusting travesty, posing as a disinterested enforcer of human rights, attacking a sovereign country which, in spite of the problematic nature of its regime, nonetheless acts as an obstacle to the unbridled assertion of the New World Order. . . . The unwritten pact of peaceful coexistence—the respect of each state's full sovereignty, that is, non-interference in internal affairs, even in the case of the grave violation of human rights—is over. However, the very first act of the new global police force usurping

the right to punish sovereign states for their wrongdoings already signals its end, its own undermining, since it immediately became clear that this universality of human rights acting as its legitimization is false, that the attacks are on selective targets in order to protect particular interests.

(See Slavoj Žižek, "Against the Double Blackmail," *New Left Review*, March/April 1999, pp. 79, 81.) What, then, is to be done? As Tariq Ali fervently tried to remind everyone in 1999, "NATO is, essentially, nothing more than an instrument to secure US hegemony in Europe and ... the world. ... The dissolution of NATO might be the first step on the long road to world peace." (See Tariq Ali, "Springtime for NATO," *New Left Review*, March/April 1999, pp. 67–68, 72.) Žižek himself was no less fervent when he declared that "the way to fight the capitalist New World Order is not by supporting local proto-fascist resistances to it, but to focus on the only serious question today: how to build *transnational* political movements and institutions strong enough to constrain seriously the unlimited rule of Capital, and to render visible and politically relevant the fact that the local fundamentalist resistances against the New World Order, from Milosevic to Le Pen and the extreme Right in Europe, are part of it" (82)? Yet we cannot forget that during the fall of 2000, Milosevic's regime was toppled, largely—some would say precisely—because of the local protest, revolt, and workers' strikes that were borne along by the Serbian masses converging on Belgrade. (See Timothy Garton Ash, "The Last Revolution," *New York Review of Books*, November 16, 2000, pp. 8–14.) Such an event, after all, begs a deeper, perhaps even more serious question: how might some version of nationalist politics—identitarian, non-identitarian, or indeed wholly otherwise—serve to buttress transnational political efforts and institution-building such that the emergent, collective, variegated agency of "transnationality" helps movements *within* nations sidestep the exclusionary pitfalls of national self-determination, the nation as absolute horizon of history, the nation as *telos*?

9. These are the words of Jacqui Alexander. See Chandra Talpade Mohanty, "Crafting Feminist Genealogies: On the Geography and Politics of Home, Nation, and Community," in *Talking Visions*, p. 486. As Mohanty rightly emphasizes, "Dialogue across differences is thus fraught with tension, competitiveness, and pain. Just as radical or critical multiculturalism cannot be the mere sum or co-existence of different cultures in a profoundly unequal colonized world multicultural feminism cannot assume the existence of a dialogue amongst feminists from different communities without specifying a just and ethical basis of such a dialogue" (486). Dignity is only one word for where we must begin in collectively crafting and recrafting the possible conditions for transnational dialogue and an ethics of multicultural translation.

10. I put "troubled legacy" in quotation marks because it is the title of chapter 6 of my book, *At the Heart of Freedom: Feminism, Sex, and Equality* (Princeton: Princeton University Press, 1998).

11. As she puts it, " . . . a just world must entail normalization; the promise of justice must attend not only to the seduction of power, but also to the anguish that knowledge must suppress difference as well as differance, that a fully just world is impossible, forever differed and different from our projections, the undecidable in the face of which we must risk the decision that we can hear the other." (See *A Critique of Postcolonial Reason*, p. 199.)

12. See Amitava Kumar, *Passport Photos* (Berkeley: University of California Press, 2000), pp. 2–14.

13. Interestingly enough, the severity of this suppression is most palpably felt within the embattled tradition of Irish literary history. As Friedrich Engels wrote in the late 1860s, "The writers of ancient Greece and Rome, and also the fathers of the Church, give very little information about Ireland. Instead there still exists an abundant native literature, in spite of the many Irish manuscripts lost in the wars of the seventeenth and eighteenth centuries. It includes poems, grammars, glossaries, annals, and other historical writings and law-books. With very few exceptions, however, this whole literature, which embraces the period at least from the eighth to the seventeenth centuries, exists only *in manuscript*. For the Irish language printing has existed only for a few years, only from the time when the language began to die out." (See Friedrich Engels, "History of Ireland," in *Ireland and the Irish Question: A Collection of Writings by Karl Marx and Friedrich Engels* [New York: International Publishers, 1972], p. 191.) In his notes and reflections on Ireland from the 1840s, Engels reveals the reasons for the Irish language "dying out": "Ir[ish] literature?—17th century, poet[ical], histor[ical], jurid[ical], then completely suppressed due to the extirpation of the Ir[ish] *literary* language—exists *only in manuscript*—publication is beginning only now—this is possible only with an oppressed people. . . . After the most savage suppression, after every attempt to exterminate them, the Irish, following a short respite, stood stronger than ever before: it seemed they drew their main strength from the very foreign garrison forced on them in order to oppress them. Within two generations, often within one, the foreigners became more Irish than the Irish . . . The more the Irish accepted the English language and forgot their own, the more Irish they became." (See Friedrich Engels, "Material for 'History of Ireland,'" in *Ireland and the Irish Question: A Collection of Writings by Karl Marx and Friedrich Engels*, pp. 210–211.)

It is not surprising, then, that young Irish radical and rebels tried throughout the nineteenth century—before and after the Famine—to reclaim their Gaelic heritage through cultural nationalism. According to Irish historian Lawrence J. McCaffrey, the emergence during the 1830s and early 1840s of a great nationalist effort to repeal the Act of Union coincided with the literary creation of the *Nation*, a journalistic and polemical outfit whose writers came to be known as Young Ireland. "They defined a nation," writes McCaffrey,

> as something beyond a political state; it was essentially a cultural and spiritual entity formed by history and tradition. They insisted that true Irish sovereignty depended on cultural independence and integrity as well as self-government. . . . The Irish language was an important Young Ireland cause . . . Young Irelanders said that Irish expressed the mind and soul of a unique, indigenous culture, and that English, a foreign tongue, represented cultural as well as political colonialism. . . . Young Ireland romanticism defined and articulated an Irish cultural nationalism that insisted on the uniqueness and brilliance of the Gaelic tradition. Maintaining that their country could never be truly sovereign unless liberated from the tentacles of British industrial materialism, Young Irelanders insisted on de-Anglicization. They worked for the preservation of the Irish language where it survived and for its restoration where it had perished as the proper vernacular of the people; the true expression of their folk-soul.

(See Lawrence J. McCaffrey, *The Irish Question: Two Centuries of Conflict* [Lexington: University of Kentucky Press, 1995], pp. 38–39, 180.)

Cultural nationalism, it is true, usually comes along with a kind of romantic invocation of a mythical past. But in the Irish case, given their particularly brutal and bloody history, is it any wonder? McCaffrey offers a vivid, almost unthinkable historical comparison we cannot but consider, if only to acknowledge its fully tragic weight: "In many ways the Irish suffered an experience during the Great Famine similar to the Jews a century later. Both groups were victims of what Albert Camus defined as ideological murder, the sacrifice of the lives of men, women, and children to economic, social, or political theories . . . Irish people dying of hunger or crowded into the bowels of an emigrant ship in the 1840s would have had scant consolation in knowing that their predicament was not the result of race hate but the price they must pay for their religion and for Britain to remain a free-enterprise economy" (57–58). No less startlingly, by 1868, even sober John Stuart Mill had no qualms admitting

> England had for ages, from motives of different degrees of unworthiness, made her yoke heavy upon Ireland. According to a well known computation, the whole land of the island had been confiscated three times over. Part had been taken to enrich powerful Englishmen and their Irish adherents; part to form the endowment of a hostile hierarchy; the rest had been given away to English and Scotch colonists, who held, and were intended to hold it, as a garrison against the Irish. The manufactures of Ireland, except the linen manufacture, which was chiefly carried on by these colonists, were deliberately crushed for the avowed purpose of making more room for those of England. The vast majority of the native Irish, all who professed the Roman Catholic religion, were . . . despoiled of all their political and most of their civil rights, and were left in existence only to plough or dig the ground, and pay rent to their task-masters. A nation which treats its subjects in this fashion cannot well expect to be loved by them.

(See John Stuart Mill, "England and Ireland," in *John Stuart Mill on Ireland*, ed. Richard Ned Lebow [Philadelphia: Institute for the Study of Human Issues, 1979], pp. 3–4.)

All this just one year after Karl Marx, in his 1867 speech on the "Irish question" to the German Workers' Educational Association in London, had summed it all up by simply declaring, "The domination over Ireland at present amounts to collecting rent for the English aristocracy." (See Karl Marx, "Record of Marx's Speech on Irish Question," in *Ireland and the Irish Question: A Collection of Writings by Karl Marx and Friedrich Engels*, p. 142.) Perhaps, then, the question arises: Why not just present the facts as they are and avoid the narrow possibilities of political and cultural rebellion presented by the nineteenth-century Irish literary public sphere? "In a society where the scrupulous collecting of social data had played its part in political control as well as in social enlightenment," writes Terry Eagleton, in what begins to sound like an answer, "the scholar's passion for fact could not be seen as entirely ideologically innocent. In any case, the facts in question were often enough rebarbative, distilling a human wretchedness belied by the impassive spirit in which they were studied . . . While scholars bend their energies to what is the case, radicals turn them to what might be." (See Terry Eagleton, *Scholars and Rebels in Nineteenth-Century Ireland* [Oxford: Blackwell Publishers, 1999], p. 131.) In view of this, Marx would seem to deserve the last word:

> Ireland is the bulwark of the *English landed aristocracy.* The exploitation of that country is not only one of the main sources of this aristocracy's material welfare; it is its greatest moral strength. It, in fact, represents the *domination of England over Ireland.* Ireland is therefore the great means by which the English aristocracy maintains *its domination in England herself.*
>
> If, on the other hand, the English army and police were to withdraw from Ireland tomorrow, you would at once have an agrarian revolution there. But the overthrow of the English aristocracy in Ireland involves as a necessary consequence its overthrow in England. And this English landed aristocracy in Ireland is an infinitely easier operation than in England herself, because in Ireland *the land question* has hitherto been the *exclusive form* of the social question, because it is a question of existence of *life and death,* for the immense majority of the Irish people, and because it is at the same time inseparable from the *national* question. This quite apart from the Irish being more passionate and revolutionary in character than the English.

(See Karl Marx, "Marx to Sigfrid Meyer and August Vogt, April 9, 1870, " in *Ireland and the Irish Question: A Collection of Writings by Karl Marx and Friedrich Engels,* pp. 292–293.)

14. It is interesting that most of the cases of the English-only movement are about the suppression of Spanish and not so much about other languages. See William Bratton and Drucilla Cornell, "Deadweight Costs and Intrinsic Wrongs of Nativism: Economics, Freedom, and Legal Suppression of Spanish," *Cornell Law Review* 84:3, March 1999, 595–695. A revised version of this essay called "Spanish Language Rights: Identification, Freedom, and the Imaginary Domain" can be found in my recent book, *Just Cause: Freedom, Identity, and Rights,* pp. 129–153.

15. For a concise theoretical discussion of *particular* and *universal* in the practice of translation, see Judith Butler, "Competing Universalities," in *Contingency, Hegemony, Universality,* pp. 162–169.

16. Among many other texts too numerous to mention, see *Eleanor Marx: Life, Work, Contacts,* ed. John Stokes (Burlington, VA: Ashgate Publishing Company), 2000; *The Reformers: Socialist Feminism,* eds. Marie Mulvey Roberts and Tamae Mizuta (London: Routledge/Thoemmes Press, 1995); *Women and Socialism/Socialism and Women: Europe Between the Two World Wars,* eds. Helmut Gruber and Pamela Graves (Oxford: Berghahn Books, 1998); Clara Zetkin, *Selected Writings,* ed. Philip S. Foner (New York: International Publishers, 1984); Friedrich Engels, *The Origin of the Family, Private Property, and the State* (New York: Pathfinder Books, 1972).

17. According to one such study, "The time spent on housework and child care in the USSR is estimated to be only 18 percent less overall than the total working time in production. Women spend more than double the time spent by men looking after home and family. . . . The fact that around 70 percent of divorce petitions are brought by women suggests a high level of dissatisfaction with the role played by men within the family. Over the last two decades, surveys have again and again painted a picture of male inactivity and female exhaustion within the home." (See

Sue Bridger, "Young Women and Perestroika," in *Women and Society in Russia and the Soviet Union*, ed. Linda Edmondson [Cambridge: Cambridge University Press, 1992], pp. 188, 197.) For a wide-ranging discussion of the "post-socialist" legacy of gender inequalities in political, everyday, and domestic life, see *Reproducing Gender: Politics, Publics, and Everyday Life After Socialism*, eds. Susan Gal and Gail Kligman (Princeton: Princeton University Press, 2000.)

18. According to an important 1979 study of Soviet and Czechoslavakian women:

> The length of paid and unpaid maternity leave and other maternity benefits (the safeguarding of jobs and seniority), together with the widespread provision of child-care facilities, are the most visible privileges women in the socialist countries enjoy over those in the capitalist ones. Every employed Soviet woman receives 16 weeks maternity leave on full pay (26 weeks in Czechoslovakia), half of it before the child is born. She may stay away from her job (without pay) for the rest of the baby's first year (3 years in Hungary, Bulgaria, and Czechoslovakia) without losing seniority or position. This is especially important for safeguarding length-of-service bonuses, pension qualifications, and the like. A new Soviet mother must also be given her regular annual paid vacation immediately after maternity leave if she so chooses, regardless of normal vacation schedules.

(See Alena Heitlinger, *Women and State Socialism: Sex Inequality in the Soviet Union and Czechoslovakia* [Montreal: McGill–Queen's University Press, 1979], p. 109.) This is quite revealing, particularly in light of the fact that "the substantial increase in the number of employed women in the West did not result from a widening of job opportunities, as it did in Eastern Europe. It merely increased the tendency for women to work in very few sectors of the economy" (98).

19. See Sonya Michel, *Children's Interests/Mothers' Rights: The Shaping of America's Child Care Policy* (New Haven: Yale University Press, 1999), pp. 192–235.

20. In fact, by the 1960s, there was an increased effort at the community level to organize for all forms of relief whose administration, distribution, and maintenance were quickly becoming part of the national ideological agenda of criminalizing immigrant families and the poor. This was happening "at a time when the southern civil rights movement had all but ended and when many activists were turning northward, drawn by the increasing turbulence of the black urban masses. This turbulence, together with the concentrations of black voters in the north, encouraged the belief that political power could be developed through mass organization. The disruptive protests which had characterized the southern movement, in short, were quickly superseded by an emphasis on the need for 'community organization' in the northern ghettos." (See Frances Fox Piven and Richard A. Cloward, *Poor People's Movements: Why They Succeed, How They Fail* [New York: Vintage, 1979], p. 265.)

21. See Susan Faludi, *Backlash: The Undeclared War Against American Women* (New York: Crown Publishers, 1991), pp. 229–280.

22. See Carol S. Robb, *Equal Value: An Ethical Approach to Economics and Sex* (Boston: Beacon Press, 1995), pp. 27–31.

23. See *Feminist Legal Theory: Readings in Law and Gender*, ed. Katharine T. Bartlett and Rosanne Kennedy (Boulder, CO: Westview Press, 1991), p. 1.

24. See Judith Stacey, *In the Name of the Family: Rethinking Family Values in the Postmodern Age* (Boston: Beacon Press, 1996).

25. See Hilary Rodham Clinton, *It Takes a Village and Other Lessons Children Teach Us* (New York: Simon & Schuster, 1996).

26. However, a very recent and important exception is Grace Chang's book, *Disposable Domestics: Immigrant Women Workers in the Global Economy* (Cambridge, MA: South End Press, 2000).

27. See interview with Zoila Rodriguez, chapter 5, pp. 130–133.

28. See the most recent work of Martha Nussbaum: *Women and Human Development: The Capabilities Approach* (Cambridge: Cambridge University Press, 2000) and *Sex and Social Justice* (Oxford: Oxford University Press, 1999).

29. See Jim Stanford, "Openness and equity: regulating labor market outcomes in a globalized economy," in *Globalization and Progressive Economic Policy* (Cambridge: Cambridge University Press, 1998), p. 245.

30. See Greg DeFreitas, *Inequality at Work* (Oxford: Oxford University Press, 1991), pp. 209–252.

31. DeFreitas, *Inequality at Work*, pp. 136–140, 231–50.

32. See Saskia Sassen, *Globalization and its Discontents* (New York: The New Press, 1998), pp. 90–91.

33. See "The Nanny Visa: The Bracero Program Revisited," in *Disposable Domestics*, pp. 93–121.

34. See "Baird Apologizes for Illegal Hiring," *New York Times*, January 20, 1993, A-12. (Cited in *Disposable Domestics*, p. 108.)

35. See "Excerpts from Chavez's Comments," *New York Times*, January 10, 2001.

36. See *Inequality at Work*, pp. 229–231.

37. Leela Fernandes, *Producing Workers* (Philadelphia: University of Pennsylvania Press, 1997), p. 88.

38. For superb historical treatments of labor, union organization, and racial discrimination, see Bruce Nelson, *Divided We Stand: American Workers and the Struggle for Black Equality* (Princeton: Princeton University Press, 2001); and Michael Keith Honey's remarkable book, *Black Workers Remember: An Oral History of Segregation, Unionism, and the Freedom Struggle* (Berkeley: University of California Press, 1999). For a truly intimate account of the union struggles of today's women and immigrant workers, see *Not Your Father's Union Movement: Inside the AFL-CIO*, ed. Jo-Ann Mort (London: Verso, 1998). And for a pathbreaking discussion of the role queer identity plays in class formation and the politics of union organizing, see *Out At Work*, eds. Kitty Krupat and Patrick McCreery (Minneapolis: University of Minnesota Press, 2000).

39. *Producing Workers*, p. 71.

40. See *Globalization and its Discontents*, p. 91.

41. See Sharryn Kamir, "Organizing the Underground Labor Force: A Conversation with Jennifer Gordon," *Regional Labor Review*, Fall 1998, p. 21.

42. See interview with Zoila Rodriguez, chapter 5, pp. 130–133.

43. See interview with Zoina Villanueva, chapter 5, p. 125.

44. Walter Benjamin, "Surrealism," in *Selected Writings: Volume 2*, ed. Michael Jennings (Cambridge, MA: Harvard University Press, 1999), p. 211. For an excellent discussion of the role profane illuminations play in sociological and anthropological research, see Avery Gordon, *Ghostly Matters* (Minneapolis: University of Minnesota Press, 1997).

45. See *A Ghost at Heart's Edge: Stories and Poems of Adoption*, eds. Susan Ito and Tina Cervin (Berkeley, CA: North Atlantic Books), 1999.

46. Spivak, it is true, does not unequivocally assert this. That is why I think it is a
matter of rhetorical suggestion or perhaps even strategic intimation on her
part. Here are her recent words on the subject: "I consider the role of the UN-
style initiative in the New World Order in the pores of this book. The wholesale Americanizing of Southern babies is another issue. Although the personal
goodwill, indeed obsession, is, in most of these cases, unquestionable, one is
also reminded of Cecil Rhodes's remark . . . 'I contend that we are the first race
in the world, and that the more of the world we inhabit the better it is for the
human race. . . . '" (See *A Critique of Postcolonial Reason*, p. 13.) Consider them
now in light of her former words: " . . . in the case of the history of imperialism, I'm much more interested in the *enabling violation* of the post-colonial situation than in finding some sort of national identity untouched by the
vicissitudes of history." (See *The Postcolonial Critic: Interviews, Strategies, Dialogues*, ed. Sarah Harasym [London: Routledge, 1990], p. 137, emphasis mine.)
    Needless to say, the colonial fantasy of Victorian liberalism—the benevo-
lent conversion of racialized others to an exclusionary sameness that not only
effaces alterity but allows the ontological privilege of "the first race" to be extended through the political domination, sabotage, and one-way inhabitation
of the world—was rather successfully maintained by such Oxford gentlemen as
Cecil Rhodes. But to forget *that* liberal vicissitude of colonialism—especially
when confronted with U.S. citizens obsessed with doing good in an unjust
world—is, for Spivak, to think "the wholesale Americanizing of Southern babies" has nothing to do with a certain imperial legacy. It is to think there is
something decidedly new about such Americanization; something bound up
with that grand, yet discontinuous, still yet tenuous historical rupture that we
all today, in one blithe way or another, call postcolonialism; something that is
in no way reducible to the counter-revolutionary repetition of the worst legiti-
mating gestures of colonialism itself. In other words, it is to enable men and
women of the global North—particularly white Anglo feminists—to violate
whatever historical protocols for an ethical relation toward the recent past and
vanishing present may be emerging in the global South.

47. As always, Spivak sees the importance of taking a certain literary refuge in deconstruction: "The situation may be described by way of the definition of
irony (akin to our general sense of irony) given by the U.S. deconstructionist
literary critic Paul de Man (1919–1983): permanent parabasis or sustained interruption from a source relating "otherwise" (allegorein = speaking otherwise) to the continuous unfolding of the main system of meaning—both the
formalization of deconstruction and, on another level of abstraction, the logic
of global development." (See *A Critique of Postcolonial Reason*, p. 430.) And yet,
recently, Spivak has also seen the merits of employing a more literal definition
of "parabasis": a step beside yet upon a ground. (See her essay, "From Haverstock Hill Flat to U.S. Classroom," in *What's Left of Theory*, eds Judith Butler,
John Guillory, and Kendall Thomas [London: Routledge, 2000], p. 31.)

48. An inseparable part of this cultural heritage and the maintenance of dual citi-
zenship is the deeply personal project of remaining in touch with the history of
Paraguay itself and the inspiring resilience of its people. "Squeezed between
South America's two biggest countries, in the center of a vast continent,
Paraguay's chief international concern has always been maintaining its indepen-
dence. . . . Throughout their history, Paraguayans have exhibited a ferocious

nationalism that translated into a willingness to fight to the death when the future of their country was at stake. This element has been a key to their survival as a nation." (See Riordan Roett and Richard Scott Sacks, *Paraguay: The Personalist Legacy* [Boulder, CO: Westview Press, 1991], p. 141.) Yet within the current global scene of late capitalism, we must also recognize that, for any transnational citizen, "the multiple-passport holder is an apt contemporary figure; he or she embodies the split between state-imposed identity and personal identity caused by political upheavals, migration, and changing global markets." (See Aihwa Ong, *Flexible Citizenship: The Cultural Logics of Transnationality* [Durham: Duke University Press, 1999], p. 2.)

As a fundamental matter of their ethicopolitical identifications with the varieties of "nation" and "nationality," the members of such a citizenry give oppositional content to the very meaning of "transnationality" itself: "*Trans* denotes both moving through space or across lines, as well as changing the nature of something. Besides suggesting new relations between nation-states and capital, transnationality also alludes to the *trans*versal, the *trans*actional, the *trans*lational, and the *trans*gressive aspects of contemporary behavior and imagination that are incited, enabled, and regulated by the changing logics of states and capitalism" (4). As Aihwa Ong astutely observes, this opposition is crucial since "traveling subjects are never free of regulations set by state power, market operations, and kinship norms.... Contrary to the popular view that sees the state in retreat everywhere before globalization, I consider state power as a positive generative force that has responded eagerly and even creatively to the challenges of global capital" (19–21). In this way, the complex relation between transnationality and citizenship cannot be captured today by the perspective of American legalism, which, at its best, simply decouples "nationality" and "functional citizenship" as a matter of constitutional law. (See Peter Schuck, "The Functionality of Citizenship," *Harvard Law Review*, vol. 110 [1997], pp. 1814–1831.)

49. Jacques Derrida has argued that Kant's distinction between dignity and price (*Wurde* and *Preis*) mirrors a higher, more "lofty" contrast between morality and the market principle. In Derrida's words, "Dignity is invested with unconditional value which should be respected as an absolute because of the moral law from which it springs. It is not subject to negotiation and stands above the market. Unlike dignity whose value cannot be measured, price is conditional, hypothetical, negotiable and expressed in figures" (70). (See Jacques Derrida, "The Principles of Pricelessness," in *What Is the Meaning of Money?*, ed. Roger-Paul Droit [Boulder, CO: Social Science Monographs, 1998], pp. 63–73). The strong assertion of "the dignity of humanity" found in *The Metaphysics Of Morals* allows this affirmative reading of Kant to be made rather easily. Derrida, however, anchors his interpretation in a few of the more memorable passages from *The Groundwork of the Metaphysics of Morals*. I quote them here in full:

> In the kingdom of ends everything either has a *price* or a *dignity*. What has a price can be replaced by something else as its *equivalent*; what on the other hand is raised above all price and therefore admits of no equivalent has a dignity.
>
> What is related to general inclinations and needs has a *market price*; that which, even without presupposing a need, conforms with a certain taste, that is, with a delight in the mere purposeless play of our mental powers, has a *fancy price*, but that which constitutes the condition under

which alone something can be an end in itself has not merely a relative worth, that is, a price, but an inner worth, that is, *dignity*.

Now, morality is the condition under which alone a rational being can be an end in itself, since only through this is it possible to be a law-giving member in the kingdom of ends. Hence morality, and humanity insofar as it is capable or morality, is that which alone has dignity (84).

(See Immanuel Kant, *Groundwork of the Metaphysics of Morals*, in *Practical Philosophy*, ed. Mary J. Gregor [Cambridge: Cambridge University Press, 1996], pp. 43–108).

Regarding the first paragraph, Derrida suggests that because "it stands higher than price, dignity belongs to a category which we qualify as 'priceless'" (71). He goes on to say, apropos the second paragraph, that "the realization of another person's dignity implies consciousness of his or her unique difference, of course, but it only becomes possible through a measure of indifference, through the neutralization of social, economic, ethnic, etc. differences. Going beyond the field of knowledge and any objective criteria, this neutralization alone ushers in the realm of dignity, that is to say the fact that each and everyone is worth as much as the next man or woman, in the sense that he or she is beyond value, that is to say, priceless" (72). "Going beyond the field of knowledge and any objective criteria"—*pace* Derrida, such movement also ushers in the realm of the aesthetic, which can never be discretely defined, for it is the "realm" of reflective judgment, the realm that is brought into being only to emerge again and yet again.

50. Martin Heidegger reminds us of the violence involved in being-interested: "To take an interest in something suggests wanting to have it for oneself as a possession, to have disposition and control over it. When we take an interest in something we put it in the context of what we intend to do with it and what we want of it. Whatever we take an interest in is always already taken, i.e., represented, with a view to something else" (see Martin Heidegger, "Kant's Doctrine of the Beautiful. Its Misinterpretation by Schopenhauer and Nietzsche," in *Nietzsche: Volume 1: The Will to Power as Art*, trans. David Farrell Krell [San Francisco: Harper & Row, 1979], p.109). The ethical way of being overwhelmed by the dignity of others radiating before us is an ethos of respect, restraint, and humility—an ethos completely without interest. It is an ethics whose conditions of aesthetic possibility may be continually renewed through the work of reflective judgment.

As I tried to suggest years ago, committing ourselves to such an ethics demands the realization that "[t]he moment of commitment is aesthetic in its orientation. It demands not only the capacity for judgment but also the ability to dream of what-is-not-yet. The ethical cannot be reduced to an aesthetic, but neither can it do without an aesthetic." (See Drucilla Cornell, "Toward a Modern/Postmodern Reconstruction of Ethics," *University of Pennsylvania Law Review*, 133:2 [January 1985], p. 380.) In his own way, Heidegger offers a wonderfully elegant formulation of this when he writes, "Whatever exacts of us the judgment 'This is beautiful' can never be an interest. That is to say, in order to find something beautiful, we must let what encounters us, purely as it is in itself, come before us in its own stature and worth. We may not take it into account in advance with a view to something else, our goals and intentions, our possible enjoyment and advantage. . . . We must freely grant to what encounters us as such its way to be; we must allow and bestow upon it what belongs to it and what it brings to us" (109).

We cannot forget, then, the great degree to which Heidegger is trying to redeem Kant's doctrine of the beautiful by amplifying the deeply ethical dimension of "disinterestedness." As Heidegger himself forcefully argues, "The misinterpretation of 'interest' leads to the erroneous opinion that with the exclusion of interest every essential relation to the object is suppressed. The opposite is the case. Precisely by means of the 'devoid of interest' the essential relation to the object comes into play. The misinterpretation fails to see that now for the first time the object comes to the fore as pure object and that such coming forward into appearance is the beautiful. The word "beautiful" means appearing in the radiance of such coming to the fore" (110).

51. Indeed, as Kant makes plain, " . . . the sublime . . . involves . . . a representation of *limitlessness* . . . the feeling of the sublime is a pleasure that only arises indirectly, being brought about by the feeling of a momentary check to the vital forces followed at once by a discharge all the more powerful, and so it is an emotion that seems to be no sport, but dead earnest in the affairs of the imagination . . . the sublime does not so much involve positive pleasure as admiration or respect . . ." (See *The Critique of Judgement*, pp. 90–91.)

52. See chapter 3, notes 43 and 44.

NOTES TO CHAPTER 6

1. See "*Las Greñudas:* Recollections on Consciousness Raising," in *Just Cause*, pp. 11–15.
2. Ruth Gertrude Kellow is Warren Kellow's daughter from his first marriage.
3. Testimony always involves a kind of haunting, which leaves us to think of pure testimony *as* impossible testimony. Jacques Derrida has tried to show this in his recent book, *Demeure: Fiction and Testimony* (Stanford: Stanford University Press, 2000). To quote Derrida:

   I can only testify, in the strict sense of the word, from the instant when no one can, in my place, testify to what I do. What I testify to is, at that very moment, my secret; it remains reserved for me. I must be able to keep secret precisely what I testify to. . . . One must oneself be present, raise one's hand, speak in the first person and in the present, and one must do this in order to testify to a present, to an indivisible moment, that is, at a certain point to a moment assembled at the tip of instantaneousness which must resist division. If that to which I testify is divisible, if the moment in which I testify is divisible, if my attestation is divisible, at that very moment it is no longer reliable, it no longer has the value of truth, reliability, or veracity that is claims absolutely. Consequently, for testimony there *must* be the instant. . . . When I commit myself to speaking the truth, I commit myself to repeating the same thing an instant later, two instants later, the next day, and for eternity, in a certain way. But the repetition carries the instant outside of itself. Consequently the instant is instantaneously, *at this very instant*, divided, destroyed by what it nonetheless makes possible—testimony. How is it that the instant makes testimony both possible and impossible at the same time? . . . In essence a testimony is always autobiographical: it tells, in the first person, the sharable and unsharable secret of what happened to me, to me, to me alone, the absolute secret of what I was in a position to live, see, hear, touch, sense, and feel. . . . The testimony testifies to nothing less than the instant of an interruption of time and history . . . (30, 33, 43, 73).

4. In chapter 4, I tried to illuminate the ethical complexity of this by discussing at length my trip to Paraguay.

5. Alicia Ostriker, "A Birthday Suite," *Green Age* (Pittsburgh: University of Pittsburgh Press, 1989), p. 20.

6. See Marta Moreno Vega, *The Altar of My Soul: The Living Traditions of Santería* (New York: Ballantine Books, 2000), pp. 265–266.

7. See *At the Heart of Freedom*, pp. 3–32.

8. For an account of the imaginary aspect of personhood, see *The Imaginary Domain: Abortion, Pornography, & Sexual Harassment*, pp. 3–27.

9. Alicia Ostriker, "Mother/Child," *The Mother/Child Papers* (Santa Monica, CA: Momentum Press, 1980), p. 23.

10. Even after I had decided that I would find some way to reconcile with them, one of the worst moments in my relationship to both of my parents was when friends told me that my parents were building an entire wing of the UCLA Business School. It is named after them—The Barbara June and Clark Cornell Wing. When my mother confessed the project to me, I was furious; but not mad enough to jeopardize our relationship. We hung in there. And the making of that wing in the UCLA Business School prevailed.

11. See *The Imaginary Domain: Abortion, Pornography, and Sexual Harassment*, pp. 3–27.

12. As I have written elsewhere, "*Per-sona*, in Latin, literally means a shining-through. A person is what shines through a mask even though the concept of the 'mask' is the usual association made with the word 'persona.' It is that which shines through. For a person to be able to shine through, she must first be able to imagine herself as whole even if she knows she can never truly succeed in becoming whole or in conceptually differentiating between the 'mask' and the 'self.' . . . A person is not something 'there' on this understanding, but a possibility, an aspiration which, because it is that, can never be fulfilled once and for all. The person is, in other words, implicated in an endless process of working through personae." (See *The Imaginary Domain: Abortion, Pornography, and Sexual Harassment*, pp. 4–5.)

13. See generally, Ronald Dworkin, *Sovereign Virtue* (Cambridge, MA: Harvard University Press, 2000); *Life's Dominion: An Argument About Abortion, Euthanasia, and Individual Freedom* (New York: Knopf, 1993); *Law's Empire* (Cambridge, MA: Harvard University Press, 1986).

14. Edwidge Danticat, *Breath, Eyes, Memory: A Novel* (New York: Vintage Books, 1998 [1994]), p. 81.

15. Danticat, *Breath, Eyes, Memory*, pp. 84–85.

16. The Tonton Macoutes were secret police under the entire Duvalier regime (Papa Doc as well as Baby Doc) in Haiti. The term is *kreyol*. It comes from local myths and fairytales in which the Tonton Macoute is a kind of werewolf creature—as in "come in when I call you or the Tonton Macoute will steal you away." It is a nickname for the police that derives from the parallel drawn between a scary creature of children's tales and the state's secret police.

17. Danticat, *Breath, Eyes, Memory*, p. 234.

18. Danticat, *Breath, Eyes, Memory*, p. 234.

19. Danticat, *Breath, Eyes, Memory*, p. 234.

BIBLIOGRAPHY

Abelin-Sas, Graciela, "The Internal Interlocutor" (unpublished manuscript on file with the author)

Abraham, Nicolas, and Torok, Maria, *The Shell and Kernel: Renewals of Psychoanalysis* (Chicago: University of Chicago Press, 1994)

Agamben, Giorgio, *Stanzas: Word and Phantasm in Western Culture* (Minneapolis: University of Minnesota Press, 1993)

Ali, Tariq, "Springtime for NATO," *New Left Review* (March/April 1999): 62–72

Arendt, Hannah, *Lectures on Kant's Political Philosophy*, ed. Ronald Beiner (Chicago: University of Chicago Press, 1982)

Ash, Timothy Garton, "The Last Revolution," *New York Review of Books* 9November 16, 2000): 8–14

Bartlett, Katharine T. and Kennedy, Rosanne, eds., *Feminist Legal Theory: Readings in Law and Gender* (Boulder, CO: Westview Press, 1991)

Benjamin, Walter, "Surrealism," in *Selected Writings: Volume 2*, ed. Michael Jennings (Cambridge, MA: Harvard University Press, 1999), pp. 207–236

Bridger, Sue, "Young Women and Perestroika," in *Women and Society in Russia and the Soviet Union*, ed. Linda Edmonson (Cambridge: Cambridge University Press, 1992), pp. 178–201

Bronfen, Elizabeth, *Over Her Dead Body: Death, Femininity, and the Aesthetic* (Manchester: Manchester University Press, 1992)

Brownmiller, Susan, *Femininity* (New York: Linden Press/Simon & Schuster, 1984)

Butler, Judith, *Antigone's Claim: Kinship Between Life and Death* (New York: Columbia University Press, 2000)

———. *Bodies That Matter: On the Discursive Limits of Sex* (New York: Routledge, 1993)

———. "Competing Universalities," in *Contingency, Hegemony, Universality: Contemporary Dialogues on the Left*, Judith Butler, Ernesto Laclau, and Slavoj Žižek (London: Verso, 2000), pp. 136–181

———. "The Future of Sexual Difference: An Interview with Drucilla Cornell and Judith Butler," *Diacritics* (Spring 1998): 19–42

———. *Gender Trouble: Feminism and the Subversion of Identity* (New York: Routledge, 1990)

———. *The Psychic Life of Power: Theories in Subjection* (Stanford: Stanford University Press, 1997)

Cantor, Norman L., *Advance Directives and the Pursuit of Death with Dignity* (Bloomington: Indiana University Press, 1993)

Cardinal, Marie, *The Words to Say It* (Cambridge, MA: Van Vactor & Goodheart, 1983)

Cascardi, Anthony, *Consequences of Enlightenment* (New York: Cambridge University Press, 1999)

Castoriadis, Cornelius, *The Imaginary Institution of Society*, trans. Kathleen Blamey (Cambridge, MA: MIT Press, 1987)

————.*The World in Fragments: Writing on Politics, Society, Psychoanalysis, and the Imagination* (Stanford: Stanford University Press, 1997)

Chang, Grace, *Disposable Domestics: Immigrant Workers in the Global Economy* (Cambridge, MA: South End Press)

Chodorow, Nancy, *The Reproduction of Mothering: Psychoanalysis and the Sociology of Gender* (Berkeley: University of California Press, 1978)

Chomsky, Noam, *A New Generation Draws the Line: Kosovo, East Timor, and the Standards Of The West* (London: Verso, 2000)

Church, John, *Pasadena Cowboy: Growing Up in Southern California and Montana 1925 to 1947* (Novato: Conover-Patterson, 1996)

Clinton, Hilary Rodham, *It Takes a Village and Other Lessons Children Teach Us* (New York: Simon & Schuster, 1996)

Coontz, Stephanie, *The Way We Never Were: American Families and the Nostalgia Trap* (New York: Basic Books, 1992)

Cornell, Drucilla, *At the Heart of Freedom: Feminism, Sex, and Equality* (Princeton: Princeton University Press, 1998)

————. *Beyond Accommodation: Ethical Feminism, Deconstruction and the Law* (Lanham: Rowman & Littlefield, 1999)

————. "The Future of Sexual Difference: An Interview with Judith Butler and Drucilla Cornell," *Diacritics* (Spring 1998)

————. *Just Cause: Freedom, Identity, and Rights* (Lanham: Rowman & Littlefield, 2000)

————. *The Imaginary Domain: Abortion, Pornography, and Sexual Harassment* (London: Routledge, 1995)

————. "Opening Remarks," in *Is Feminist Philosophy Philosophy*, ed. Emanuela Bianchi (Evanston: Northwestern University Press, 1999), pp. 3–9

————. *The Philosophy of the Limit* (London: Routledge, 1992)

————. "Rethinking the Beyond of the Real," in *Levinas and Lacan: The Missed Encounter*, ed. Sarah Harasym (Albany: State University of New York Press, 1998), pp. 139–181

————. "Toward a Modern/Postmodern Reconstruction of Ethics," *University of Pennsylvania Law Review*, vol. 133, no. 2 (January 1985): 291–380

————. *Transformations: Recollective Imagination and Sexual Difference* (London: Routledge, 1993)

————. "Who's Afraid of Disorderly Conduct" (forthcoming)

Danticat, Edwidge, *Breath, Eyes, Memory: A Novel* (New York: Vintage Books, 1998)

de Beauvoir, Simone, *The Second Sex*, trans. H. M. Parshley (New York: Vintage Books, 1974)

DeFreitas, Greg, *Inequality at Work* (Oxford: Oxford University Press, 1991)

Derrida, Jacques, *Demeure: Fiction and Testimony* (Stanford: Stanford University Press, 2000)

————. *Memoires: For Paul De Man* (New York: Columbia University Press, 1986)

————. "Opening Remarks," in *Is Feminist Philosophy Philosophy*, ed. Emanuela Bianchi (Evanston: Northwestern University Press, 1999), pp. 13–14

————. "The Principles of Pricelessness," in *What Is the Meaning of Money?*, ed. Roger-Paul Droit (Boulder, CO: Social Science Monographs, 1998), pp. 63–73

Dworkin, Ronald, "Assisted Suicide: The Philosophers' Brief," Ronald Dworkin, Thomas Nagel, Robert Nozick, John Rawls, Thomas Scanlon, and Judith Jarvis Thomson, *New York Review of Books* (March 27, 1997)

————. *Law's Empire* (Cambridge, MA: Harvard University Press, 1986)

————. *Life's Dominion: An Argument About Abortion, Euthanasia, and Individual Freedom* (New York: Knopf, 1993)

————. *A Matter of Principle* (Cambridge, MA: Harvard University Press, 1985)

————. *Sovereign Virtue* (Cambridge, MA: Harvard University Press, 2000)

Eagleton, Terry, *The Ideology of the Aesthetic* (Oxford: Blackwell Press, 1990)

————. *Scholars and Rebels in Nineteenth-Century Ireland* (Oxford: Blackwell Press, 1999)

Ehrenberg, Darlene Bregman, *The Intimate Edge: Extending the Reach of Psychoanalytic Interaction* (New York: W.W. Norton & Company, 1992)

Engels, Friedrich, *The Origin of the Family, Private Property, and the State* (New York: Pathfinder Books, 1972)

Faludi, Susan, *Backlash: The Undeclared War Against American Women* (New York: Crown Publishers, 1991)

Fernandes, Leela, *Producing Workers* (Philadelphia: University of Pennsylvania Press, 1997)

Gal, Susan and Gail Kligman, eds., *Reproducing Gender: Politics, Publics, and Everyday Life After Socialism* (Princeton: Princeton University Press, 2000)

Gordon, Avery, *Ghostly Matters* (Minneapolis: University of Minnesota Press, 1997)

Grewal, Inderpal, "On the New Global Feminism and the Family of Nations: Dilemmas of Transnational Feminist Practice," in *Talking Visions: Multicultural Feminism in a Transnational Age*, ed. Ella Shohat (Cambridge, MA: MIT Press, 1998), pp. 501–530

Grimm, Nicole L., "The North American Free Trade Agreement on Labor Cooperation and Its Effects on Women in Mexican Maquiladoras," *American University Law Review* 48 (October 1998): 179–227

Grosz, Elizabeth, *Jacques Lacan: A Feminist Introduction* (New York: Routledge, 1990)

Gruber, Helmut and Pamela Graves, *Women and Socialism/Socialism and Women: Europe Between the Two World Wars* (Oxford: Berghahn Books, 1998)

Gurewich, Judith Feher, "Is the Prohibition of Incest A Law?" (unpublished manuscript on file with the author)

————. "The *Jouissance* of the Other and the Prohibition of Incest: A Lacanian Perspective" (unpublished manuscript on file with the author)

————. "The Subversive Value of Symbolic Castration: The Case of Desdemona," in *JPCS: Journal for the Psychoanalysis of Culture & Society* 2 (Fall 1997): 61–66

Hampton, Jean, "Feminist Contractarianism," in *A Mind of My Own: Feminist Essays on Reason and Objectivity*, eds. Louise M. Antony and Charlotte Wilt (Boulder, CO: Westview Press, 1993)

Heidegger, Martin, *Being and Time* (New York: Harper & Row, 1962)

————. "Kant's Doctrine of the Beautiful. Its Misinterpretation by Schopenhauer and Nietzsche," in *Nietzsche Volume 1: The Will to Power as Art*, trans. David Farrell Krell (San Francisco: Harper & Row, 1979), pp. 107–114

————. *Kant and the Problem of Metaphysics* (Bloomington: Indiana University Press, 1997)

Heitlinger, Alena, *Women and State Socialism: Sex Inequality in the Soviet Union and Czechoslovakia* (Montreal: McGill - Queen's University Press, 1979)

Honey, Michael Keith, *Black Workers Remember: An Oral History of Segregation, Unionism, and the Freedom Struggle* (Berkeley: University of California Press, 1999)

hooks, bell, *All about Love: new visions* (New York: William Marrow, 2000)

Humphrey, Derek and Ann Wickett *The Right to Die: Understanding Euthanasia* (New York: Harper & Row, 1986)

Hurston, Zora Neale, *Their Eyes Were Watching God* (New York: HarperCollins Publishers, 1990)

Ito, Susan, and Tina Cervin, eds., *A Ghost at Heart's Edge: Stories and Poems of Adoption* (Berkeley: North Atlantic Books, 1999)

Kasmir, Sharryn, "Organizing the Underground Labor Force: A Conversation with Jennifer Gordon of the Workplace Project," *Regional Labor Review* (Fall 1998): 17–23

Kant, Immanuel, *The Critique of Judgment*, trans. James Meredith Smith (Oxford: Oxford University Press, 1952)

———. *The Critique of Pure Reason*, trans. James Meredith Smith (New York: St. Martin's Press, 1929)

———. *Ethical Philosophy* (Indianapolis: Hackett Publishing Company, 1994)

———. *The Groundwork of the Metaphysics of Morals*, in *Practical Philosophy*, ed. Mary J. Gregor, The Cambridge Edition of the Works of Immanuel Kant (Cambridge: Cambridge University Press, 1996), pp. 48–108

———. *The Metaphysics of Morals*, in *Practical Philosophy*, ed. Mary J. Gregor, The Cambridge Edition of the Works of Immanuel Kant (Cambridge: Cambridge University Press 1996), pp. 365–506

———. *The Metaphysics of Morals*, trans. Mary J. Gregor (Cambridge: Cambridge University Press, 1993)

Korsgaard, Christine M., *Creating the Kingdom of Ends* (Cambridge: Cambridge University Press, 1996)

Krupat, Kitty, and Patrick McCreery, eds., *Out At Work* (Minneapolis: University of Minnesota Press, 2000)

Kumar, Amitava, *Passport Photos* (Berkeley: University of California Press, 2000)

Lacan, Jacques, *The Four Fundamental Concepts of Psychoanalysis* (New York: W.W. Norton & Company, 1998)

La Capra, Dominick, *Rethinking Intellectual History: Texts, Contexts, Language* (Ithaca: Cornell University Press, 1983)

Lacey, Nicola, *Unspeakable Subjects: Feminist Essays in Legal and Social Theory* (Oxford: Hart Publishing, 1998)

Laclau, Ernesto, "Identity and Hegemony: The Role of Universality in the Constitution of Political Logics," in *Contingency, Hegemony, Universality: Contemporary Dialogues on the Left*, ed. Judith Butler, Ernesto Laclau, and Slavoj Žižek (London: Verso, 2000), pp. 44–89

Mackinnon, Catharine A., *In Harm's Way: The Pornography Civil Rights Hearings* (Cambridge, MA: Harvard University Press, 1997)

Marx, Karl and Friedrich Engels, *Ireland and the Irish Question* (New York: International Publishers, 1972)

McCaffrey, Lawrence J., *The Irish Question: Two Centuries of Conflict* (Lexington: University of Kentucky, 1995)

Mehata, Uday Singh, *Liberalism and Empire: A Study in Nineteenth Century British Political Thought* (Chicago: University Press of Chicago, 1999)

Michel, Sonya, *Children's Interests/Mothers' Rights: The Shaping of America's Child Care Policy* (New Haven: Yale University Press, 1999)

Minh-ha, Trinh T., *Woman, Native, Other: Writing Postcoloniality and Feminism* (Bloomington: Indiana University Press)

Mohanty, Chandra Talpade, "Crafting Feminist Genealogies: On the Geography and Politics of Home, Nation, and Community," in *Talking Visions: Multicultural Feminism in a Transnational Age*, ed. Ella Shohat (Cambridge, MA: MIT Press, 1998), pp. 485–500

Momeyer, Richard W., *Confronting Death* (Bloomington: Indiana University Press, 1988)

Morrison, Toni, *Beloved* (New York: Penguin, 1987)

———. *The Nobel Lecture in Literature* (New York: Knopf, 2000)

———. *Sula* (New York: Penguin, 1973)

Mort, Jo-Ann, ed., *Not Your Father's Union Movement: Inside The AFL-CIO* (London: Verso, 1998)

Nancy, Jean-Luc, *The Experience of Freedom* (Stanford: Stanford University Press, 1993)

Nelson, Bruce, *American Workers and the Struggle for Black Equality* (Princeton: Princeton University Press, 2001)

Nussbaum, Martha, *Sex and Social Justice* (Oxford: Oxford University Press, 1999)

———. *Women, Culture, and Development: A Study of Human Capabilities*, eds. Martha C. Nussbaum and Jonathan Glover (New York: Oxford University Press, 1995)

———. *Women and Human Development: The Capabilities Approach* (Cambridge: Cambridge University Press, 2000)

Oliver, Kelly, *Subjectivity Without Subjects: Abject Fathers to Desiring Mothers* (Lanham: Rowman & Littlefield, 1998)

Ong, Aihwa, *Flexible Citizenship: The Cultural Logics of Transnationality* (Durham & London: Duke University Press, 1999)

Ostriker, Alicia, *The Imaginary Lover* (Pittsburgh: University of Pittsburgh Press, 1986)

———. *Green Age* (Pittsburgh: University of Pittsburgh Press, 1989)

———. *The Mother/Child Papers* (Santa Monica, CA: Momentum Press, 1980)

Pierce, Charles Sanders, "Prolegomena to an Apology for Pragmaticism," in *Pierce on Signs*, ed. John Hoopes (Chapel Hill: University of North Carolina Press, 1991), pp. 249–252

Piven, Frances Fox and Richard A. Cloward, *Poor People's Movements: Why They Succeed, How They Fail* (New York: Vintage, 1979)

Rawls, John, *Collected Papers*, ed. Samuel Freeman (Cambridge, MA: Harvard University Press, 1999)

———. *Political Liberalism* (Cambridge, MA: Harvard University Press, 1993)

———. *A Theory of Justice* (Cambridge, MA: Harvard University Press, 1971)

Riviere, Joan, "Womanliness as a Masquerade," in *Formations of Fantasy*, ed. Victor Burgin, James Donald, and Cora Kaplan (London: Methuen, 1986)

Robb, Carol S., *Equal Value: An Ethical Approach to Economics and Sex* (Boston: Beacon Press, 1995)

Roberts, Marie Mulvey and Tamae Mizuta, eds., *The Reformers: Socialist Feminism* (London: Routledge/Thoemmes Press, 1995)

Roett, Riordan and Richard Scott Sacks, *Paraguay: The Personalist Legacy* (Boulder, CO: Westview Press, 1991)

Rorty, Richard, *Philosophy and the Mirror of Nature* (Princeton: Princeton University Press, 1980)

Ruddick, Sara, "Injustice in Families: Assault and Domination," in *Justice and Care: Essential Readings in Feminist Ethics*, ed. Virginia Held (Boulder, CO: Westview Press, 1995), pp. 203–223

Said, Edward, "Protecting the Kosovars?," *New Left Review* (March/April 1999): 73–75

Sandel, Michael J., *Liberalism and the Limits of Justice* (Cambridge: Cambridge University Press, 1998)

Sassen, Saskia, *Globalization and Its Discontents* (New York: The New Press, 1998)

Schiller, Friedrich, *Essays*, ed. Walter Hinderer and Daniel O. Dahlstrom, trans. Daniel O. Dahlstrom (New York: Continuum Publishing Company, 1998)

Schuck, Peter, "The Functionality of Citizenship," *Harvard Law Review*, vol. 110 (1997): 1814–1831

Scott, Joan Wallach, *Only Paradoxes to Offer: French Feminists and the Rights of Man* (Cambridge, MA: Harvard University Press, 1996)

Shahar, Shulamith, *The Fourth Estate: A History of Women in the Middle Ages*, trans. Chaya Galai (London: Methuen & Co., 1983)

Shepherdson, Charles, "The Epoch of the Body: On the Domain of Psychoanalysis," in *Perspectives on Embodiment: The Intersections of Nature and Culture*, ed. Gail Weiss and Honi Haber (London: Routledge, 1999), pp. 183–211

———. *Vital Signs: Nature, Culture, Psychoanalysis* (New York: Routledge, 2000)

Silverman, Kaja, *World Spectators* (Stanford: Stanford University Press, 2000)

Singer, Peter, "Unsanctifying Life," in *Ethical Issues Relating to Life and Death*, ed. John Ladd (Oxford: Oxford University Press, 1979), pp. 41–61

Spivak, Gayatri Chakravorty, *A Critique of Postcolonial Reason: Toward a History of the Vanishing Present* (Cambridge, MA: Harvard University Press, 1999)

———. "From Haverstock Hill Flat to U.S. Classroom," in *What's Left of Theory*, eds. Judith Butler, John Guillory, and Kendall Thomas (London: Routledge, 2000)

———. *In Other Worlds* (New York and London: Methuen, 1987)

———. *The Postcolonial Critic: Interviews, Strategies, Dialogues*, ed. Sarah Harasym (London: Routledge, 1990)

Stacey, Judith, *In the Name of the Family: Rethinking Family Values in the Postmodern Age* (Boston: Beacon Press, 1996)

Stanford, Jim, "Openness and Equity: Regulating Labor Market Outcomes in a Globalized Economy," in *Globalization and Progressive Economic Policy*, ed. Dean Baker, Gerald Epstein, and Robert Pollin (New York: Cambridge University Press, 1998), pp. 245–270

Stokes, John, ed., *Eleanor Marx: Life, Work, Contacts* (Burlington, VA: Ashgate Publishing Company, 2000)

Tobias, Sheila, *Faces of Feminism: An Activist's Reflections on the Women's Movement* (Boulder, CO: Westview Press, 1997)

Tooley, Michael, "Decisions to Terminate Life and The Concept of the Person," in *Ethical Issues Relating to Life and Death*, ed. John Ladd (Oxford: Oxford University Press, 1979), pp. 62–93

Vega, Marta Moreno, *The Altar of My Soul: The Living Traditions of Santería* (New York: Ballantine Books, 2000)

Verhaeghe, Paul, "The Collapse of the Function of the Father and Its Effect on Gender Roles" (unpublished manuscript on file with author)

———. *Does the Woman Exist?: From Freud's Hysteria to Lacan's Feminine*, trans. Marc Du Ry (New York: Other Press, 1998)

Walker, Margaret Urban, "Moral Understandings: Alternative 'Epistemology' For a Feminist Ethics," in *Justice and Care: Essential Readings in Feminist Ethics*, ed. Virginia Held (Boulder, CO: Westview Press, 1995), pp. 139–52

Walzer, Michael, "Feed the Face," *The New Republic* (June 9, 1997)

Zetkin, Clara, *Selected Writings*, ed. Philip S. Foner (New York: International Publishers, 1984)

Žižek, Slavoj, "Against the Double Blackmail," *New Left Review* (March/April 1999): 76–82

———. "Melancholy and the Act," *Critical Inquiry*, vol. 26. no. 4 (Summer 2000): 657–681

INDEX

Printed in the United States
by Bookmasters

Printed in the United States
By Bookmasters